The Yamaha XS650 Engine

Including the Electric System

Hans J. Pahl

Copyright © 2017 Hans Joachim Pahl
All rights reserved.
ISBN-10: 1544270631
ISBN-13: 978-1544270630

DEDICATION

If you own a Yamaha XS650 – a motorcycle which is no longer produced for more than 30 years now - you should be able to do maintenance and repair work yourself. This book is intended to help those repairing their own engines, who are no trained mechanics, but who have just basic knowledge. It is meant to close the gap, between those, whose job it is to work on a variety of engines every day in a workshop, and those who work in other jobs.

As a supplement to the original workshop manual, this book provides a very detailed description of the function of the engine, which is based on more than 400 color photos. Anyone who has the necessary equipment and basic knowledge should be able to perform maintenance and repair-work on the XS650 engine.

Problems in the electrical system are actually easy to fix. If you don't want to do it by trial and error, this book helps you to understand the electric system of the XS650. You will find a detailed description of the function and how to test all components of the power supply on 35 pages with 38 color photos and graphics. In addition a highly simplified wiring harness, which contains all the necessary functions, is described.

New spare parts are rarely available. So a whole chapter with 34 color photos is dedicated to the description of typical damages. It will help you to recognize the beginning of wear and to decide wether a part should be replaced or not.

ACKNOWLEDGMENTS

I would like to thank Reiner Althaus from Wuppertal, in whose workshop I took the photos in the chapters 3 and 6. Furthermore I would like to thank Udo Land from Bochum, with whom I worked together when making the first edition in German language.

Contents

1 Introduction 1
2 Description of the funktion 2
   2.1 Engine..6
      2.1.1 Crankschaft, pistons 7
      2.1.2 The Cylinders..11
      2.1.3 Timing chain, camshaft...12
      2.1.4 Valve actuation 16
   2.2 Power Transmission 18
      2.2.1 Primary drive...18
      2.2.2 The Clutch 18
      2.2.3 Clutch actuation 22
      2.2.4 Gear box 23
      2.2.5 Transmission ratios 24
      2.2.6 Structure 25
      2.2.7 Funktion, drive train 26
      2.2.8 Shifting mechanism 29
      2.2.9 Kickstarter...35
   2.3 Oil circuit 36
      2.3.1 The oilfilter..37
      2.3.2 The oil pump...39
      2.3.3 Lubrication points of the engine 41
      2.3.4 The lubrication of the transmission...45
3 Components 46
   3.1 Images 46
      3.1.1 The engine..47
      3.1.2 Transmission 50
   3.2 Spare parts situation..51
   3.3 Dealers 53
4 Tools 57
5 Removing the engine from the frame..63
6 Disassambling the engine 66
   6.1 Disassembling the cylinder head 68
   6.2 Dismantling the pistons 72
   6.3 Dismantling the generator 73
   6.4 Dismantling the clutch...74
   6.5 Disconnecting the engine housing halves 77
   6.6 Removing the crankshaft and the transmission 79
7 Assembling the engine...81
   7.1 Shifting mechanism, gear box and kickstarter...................................83
   7.2 Closing the engine housing halves..87
   7.3 Installation of the shift shaft and the clutch 90
      7.3.1 The shift shaft 90
      7.3.2 The clutch 91
   7.4 Install the right side crankcase cover 95

7.5	Clutch-pushrod and the drive sprocket		97
7.6	The generator		98
7.7	The pistons and cylinders		100
7.8	Cylinderhead and camshaft		102
	7.8.1	Riveting the timing chain	105
	7.8.2	The cylinder head cover	106
7.9	The timing chain tensioner		107
7.10	Concluding work		109
8	**Typical damage**		**110**
8.1	Engine		110
	8.1.1	Crankshaft	110
	8.1.2	Pistons	110
	8.1.3	Valve drive, timing chain, tensioning rails	111
	8.1.4	Camshaft	112
	8.1.5	Valves	113
8.2	The clutch		114
8.3	Transmission		116
	8.3.1	Gear wheels	116
	8.3.2	Shifting claws	117
	8.3.3	Shift forks	118
8.4	Oil circuit		118
	8.4.1	Oilfilter	118
	8.4.2	Oil pump	119
9	**Electric system**		**121**
9.1	Simplified electrical system		122
	9.1.1	Consumer circuit	122
	9.1.2	Charging circuit	127
9.2	Funktion of the charging circuit		128
	9.2.1	The generator (alternator)	128
	9.2.2	The rectifier	133
	9.2.3	The voltage regulator	135
	9.2.4	The ignition lock	138
9.3	Check the components of the charging circuit		139
	9.3.1	Measuring instruments	140
	9.3.2	Check for leakege currents	142
	9.3.3	Checking the charging current	142
	9.3.4	Testing the rotor	144
	9.3.5	Testing the stator	145
	9.3.6	Testing the carbon brushes	146
	9.3.7	Testing the rectifier	147
	9.3.8	Testing the voltage regulator	148
	9.3.9	Testing the ignition lock	150
9.4	The ignition circuit		151
9.5	Adjusting the ignition timing		156

1 Introduction

The original workshop manual contains in principle all the information required for maintenance and repair work, but only a few black and white illustrations. This book is therefore not a manual of its own, but it is intended to complement the original manual, which is online available. Most of the information in the original workshop manual, such as technical data, the differences between the different versions, wear and so on, are not repeated here since it would make this book unnecessarily extensive and thus expensive. Before reading in this book you should download the original manual, because I often refer to it.

The original workshop manual is based on the fact that you are working on an engine in its original state. If more than 30 years after these engines are no longer built, such an engine is opened, one can hardly expect to find it in its original state. Some assemblies, e.g. the clutch or the timing chain tensioner have been changed several times during the production time. If these parts have been removed from an engine that has been running so far, then you can install them again. If, however, you are assembling parts of different origin, for example parts of the timing chain tensioner, damage can occur if the timing chain is either too tight or not tensioned at all in case that you follow the description in the original workshop manual.

Before you begin working on an engine, you should be aware of the function of all assemblies, and always check - even during assembly - whether the assembly is working properly.

Since it is very important to fully understand the functions of the modules of the engine and the gearbox, the functionality of all assemblies is described in chapter "2 Description of the funktion" in detail and is illustrated with many graphics and color photos. In addition the photos, I also used explosions drawings from the spare parts list and detail photos of a cutaway model, used by the German importer for training purposes, to illustrate the design of the assemblies.

First of all, it is necessary to overcome the inhibition threshold to open an unknown engine.Therefore in chapter "3 Components", the individual components of the engine are shown in the manner of an explosive drawing so that one can become familiar with the actual appearance of all components. In addition, the development of the spare parts situation is described and some customizers and spare parts dealers are introduced.

The required tools are described in chapter "4 Tools". Before starting repairwork, you should be aware which tools are necessary. No expensive special tools are required to repair the XS650 engine. For example, the piston pins can be removed without tools by heating the piston. The piston rings can also be removed with some skill without a special tool and the pistons can be pushed into the cylinder liners without piston ring tensioning straps. In some cases, however, to make some aids is much faster than the actual repair without these tools. In any case, an engine stand is always helpful, with which you can always bring the engine into the most favorable position to work.

In chapter "5 Removing the engine from the frame" it is shown, how the engine can be removed from the frame by one person and without force exertion by means of a device developed by Reiner Althaus from Wuppertal.

Chapter "6 Disassembling the engine" then describes by means of photographs how the engine is disassembled into its parts. Although the assembly of the engine is in principle carried out in the reverse order, it is described in detail in Chapter "7 Assembling the engine" on the basis of photos.

In chapter "8 Typical damage", the typical damage to the XS 650 engine, which can result from normal wear and tear, lack of maintenance or faulty operation, is shown. This chapter, of course, only describes damages, which are typical for the XS 650 engine. I have therefore not been concerned with the most frequent engine damage, the piston seizure, since this and its causes are described in detail, in printed publications of piston manufacturers.

If you do not want to just exchange parts but want to understand the electrics of the XS 650, you will find in chapter "9 Electric system" a detailed description of the function and how to test the individual components of the power supply and the consumers. In addition, the preparation of a highly simplified wiring harness, which, however, contains all the necessary functions, is described.

2 Description of the funktion

In Germany the engine of a "big bike" was a two - cylinder boxer engine, as it was known in the Zündapp KS 601 model, the so - called "green elephant" and as it was built by BMW in many models. At the same time in England, the country with the leading motorcycle industry in the sixties, it was the parallel-twin, built by the three major brands Triumph, BSA and Norton, but also by some less well-known companies such as Royal Enfield.

A parallel-twin, as the name suggests, is a two-cylinder engine whose two pistons move up and down simultaneously. This has the advantage of a high torque at low engine speeds, which is bought with the disadvantage of strong vibrations over the entire speed range. At low speeds, these vibrations are still perceived by some drivers as "sturdy", while they can have a disturbing effect in higher speed ranges. Other parallel-twins, such as the Kawasaki Z 750, had compensating shafts to reduce the vibrations to an acceptable level.

Cutaway model of the engine
Figure 2-1

In the sixties the change of the motorcycle from the utility vehicle, as it was only used by police officers, to a leisure vehicle took place. It was not so much a matter of longevity and reliability but more a matter of performance and technical features with which one could impress one's friends. Four cylinders with as many exhaust pipes were standard. The traffic density was not as high as today and in some countries there were still no general speed restrictions on country roads so that the existing power could be used to its full extend.

The development and the design of the Yamaha XS 650, which was presented at the Tokyo Motor Show in 1969, took place towards the end of this period, when the motorcycle changed from the utility vehicle to the purely leisure vehicle. The XS 650 or XS 1 and XS 2, as the first models were called, therefore tended more towards the utility vehicle. The engine was deliberately kept simple, without a balancing shaft. In the first model series, there was even no electric starter. From the outside it was the classic English motorcycle engine, but inside there were significant differences:

The engine and the transmission are housed in a common housing which, in contrast to the English parallel-twins, is horizontally divided. This design ensures that the engines don't leak oil at least during standstill and it offers advantages in production. The overhead camshaft is driven by a chain, and with a stroke/bore ratio with 74/75 it is no longer a so-called long stroke engine. In comparison to a long stroke engine such as the Triumph Bonneville, there is of course a little bit of torque lost, which however benefits the longevity of the engine.

Vehicle drives consist of the following functional groups: the engine itself, the connection between the engine and the transmission (primary drive and clutch), the transmission for adapting the engine speed to the driving speed and the driving resistances, and the secondary drive. In the case of passenger cars and trucks, these are separate assemblies, which are often purchased as separate units from various suppliers. In modern motorcycle drives - and in this sense the engine of the XS 650 is a modern engine - all components of the drive train are located in a common housing. Such housings are usually divided horizontally, the crankshaft and the transmission shafts being located in the parting gap between the two housing halves.

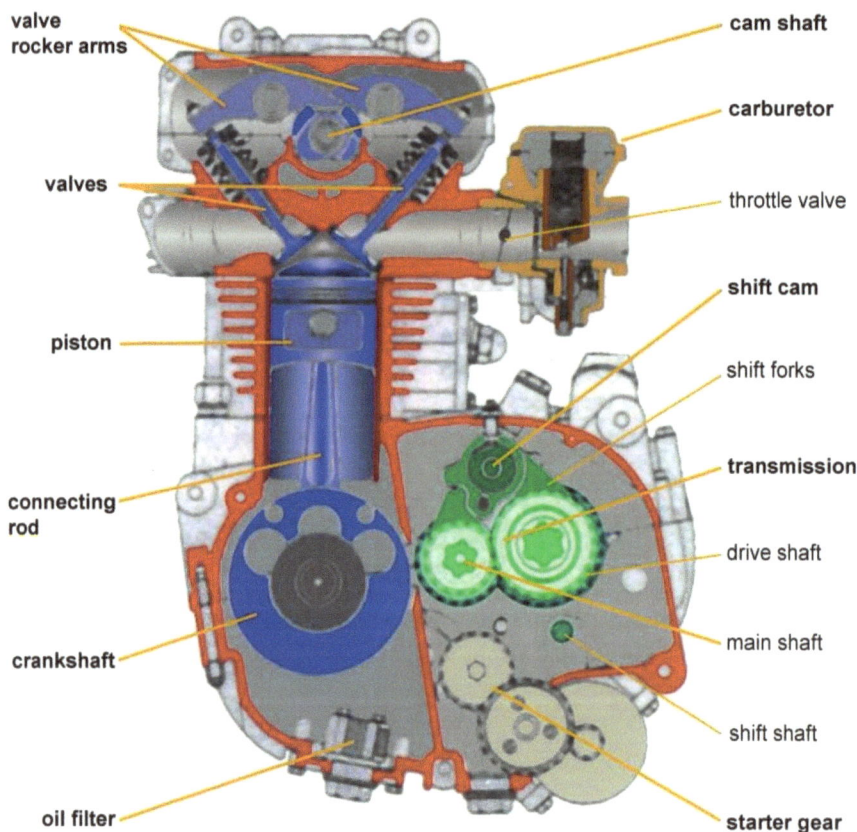

Schematic drawing of the components of the engine
Figure 2-2

Figure 2-2 shows schematically the components of the XS 650 engine. In the following, the term "engine" is used for the whole unit as well as only for the unit consisting of the crankshaft and the camshaft, the pistons and the valve train. The actual engine, is marked by blue color. The transmission and the gear shifting mechanism are highlighted by a green color marking.

The XS 650 engine is basicly an engine with a very simple design, with the exception of the electric starter. For example, there is no balancing shaft to mitigate the unavoidable vibrations of a parallel-twin. Contrary to the fashion in the 70's there is also only one overhead camshaft. As a concession to the taste of the time, an electric starter has been retrofitted, which is incompatible with the otherwise clear and purpose-built design of the engine with an elaborate execution of its transmission parts. While the whole design of the engine is designed for durability and to be easy to repair, the electric starter's transmission parts wear out very quickly and also affect other components due to the resulting metal abrasion.

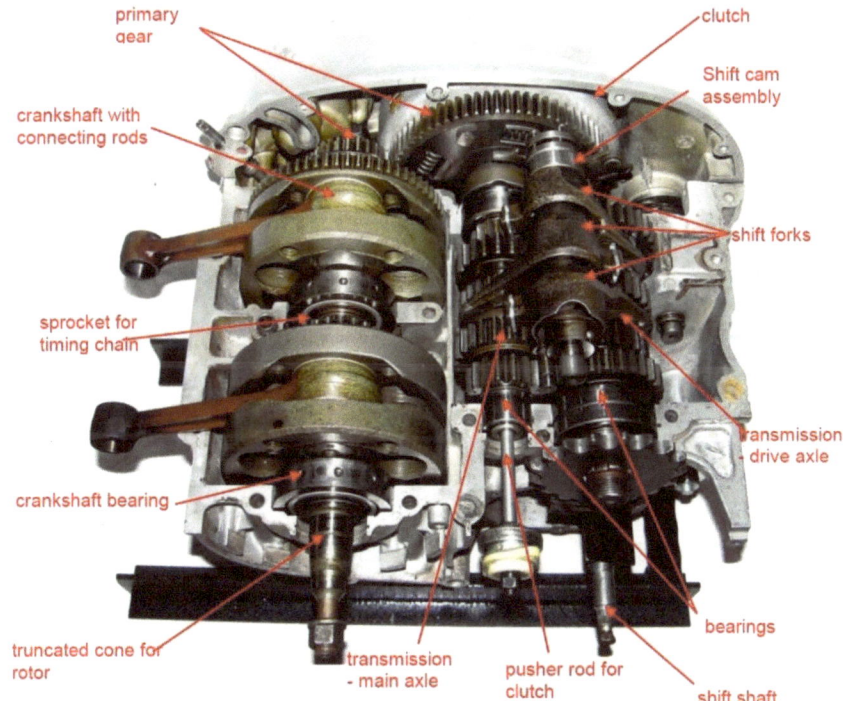

Crankshaft, primary gear, clutch, shifting mechanism and gearbox
Figure 2-3

As far as the durability and repair possibilities are concerned, the XS 650 engine is very elaborately designed. Thus, all rotating parts except the kickstarter shaft are mounted in rolling bearings. A mileage of 150,000 miles and significantly more without extensive repairs is possible with regular maintenance. An overhaul of the engine is worthwhile in any case because all parts subject to wear can be replaced and afterwards a new condition is restored.

Figure 2-3 on page 5 shows the crankshaft, the primary drive with the clutch, and the gearbox with the shifting mechanism in the lower engine housing.

The crankshaft with in total four main bearings is parallel to the two transmission shafts, each with one fixed bearing and one floating bearing. The primary drive and the clutch as well as the oil pump are located on the right side of the engine. The end of the shift shaft, which carries the shift lever, is - unlike with English motorcycles - located on the left side of the engine. The actual shifting mechanism is located on the right side of the engine. Centrally above the transmission shafts, the shift cam assembly with the shift forks is mounted in a needle bearing on the left engine side and a ball bearing on the right engine side in the upper half of the engine housing.

2.1 Engine

The components of the "engine" are the crankshaft with four main bearings, the connecting rods and the pistons, the valves and the valve train. I do not mention the electric starter, because it should be used as rarely as possible. When the ignition is set correctly, the engine will easily start up with the kickstarter. The electric starter was subsequently designed and integrated into the engine. Its actuation causes considerable wear and tear.

Crankshaft with connecting rod and piston, valve rocker arm and right intake valve
Figure 2-4

2.1.1 Crankschaft, pistons

The crankshaft is a so-called "built crankshaft", which consists of individual parts and can be dismantled. The individual parts of the crankshaft are shown on page 22 in the original manual. On the following figures 2-5 and 2-6, the crankshaft is shown by means of an explosion drawing from the spare parts list and in the assembled state by means of a photo. A dismantling and reassembling of the crankshaft is only possible with a press, as it is available in engine service workshops. Since dismantling and assembling of the crankshaft is not possible by the means which are available in a conventional workshop, it is not described here.

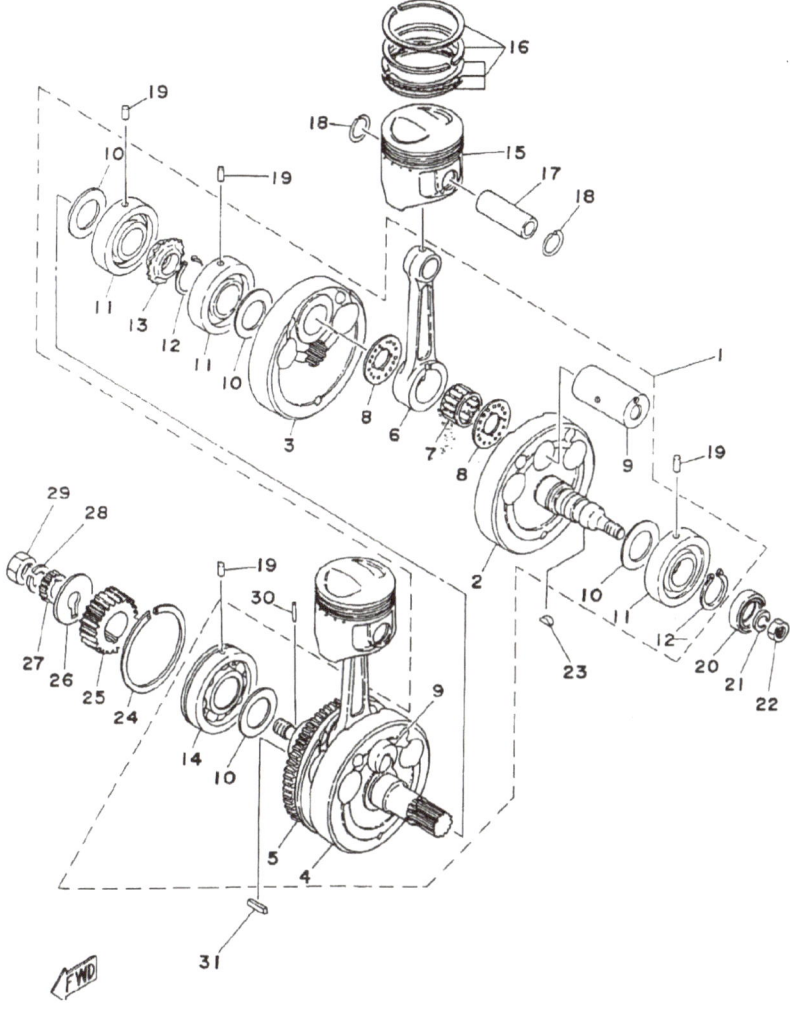

Crankshaft and piston
Figure 2-5

The crankshaft is mounted in four main bearings, of which the right bearing on the drive-side is a ball bearing, which is axially secured in the housing with a securing ring.

The other three main bearings are axially moveable roller bearings, the outer rings of which are positioned by pins so that the oil bores of the bearing's outer rings are aligned with the oil bores in the lower housing half.

Crankshaft
Figure 2-6

The figures 2-7 and 2-8 show the two central main bearings of the crankshaft with the drive pinion of the timing chain in the middle.

The rollers of the bearings are guided in cages and are not moveable in the axial direction against the inner rings of the bearings while the outer rings on the rollers can be moved in the axial direction in order to compensate for the thermal expansion of the crankshaft.

Figure 2-7

Generator side of the crankshaft wit floating bearing
Figure 2-8

Figure 2-8 shows the left main bearing of the crankshaft with the outer ring removed and a part of the tapered crankshaft stub for accommodating the generator rotor. The inner ring of the bearing is fixed in the axial direction by a retaining ring. Figure 2-8 also shows the pressed-in crank pins of the left-hand connecting rod, as well as two relief holes which compensate for the unbalance caused by the weight of the crankpin. Figure 2-9 shows the drive side of the crankshaft with the gearwheel of the electric starter.

Drive side of the crankshaft and right main
bearing (fixed bearing)
Figure 2-9

The wear and tear on the front edges of the teeth caused by being grinded during the start procedure is clearly visible here. With item 2, the pinion of the primary drive is shown in Figure 2-9, and the drive pinion of the oil pump is shown with item 3. On the left, the right main bearing of the crankshaft, designed as a ball bearing, can be seen with a circlip for axial fixing in the housing (arrow marking).

The connecting rod bearings are designed as needle bearings with thrust washers between the side surfaces of the connecting rod and the flywheels. There is no separate bearing in the connecting rod eye, i.e. the piston pin is mounted directly in the connecting rod eye. Further details on the crankshaft and wear dimensions as well as the description of the measurement can be found in the original workshop manual on page 27.

Figure 2-10 shows the pistons with two compression rings and an oil wiper.

Piston with pin and safety clip
Figure 2-10

The upper of the two compression rings is shallower and narrower than the lower one. Near the gap the letter "B" is stamped, which marks the upper side. In addition, there is an identification in the form of a two-digit combination of digits.

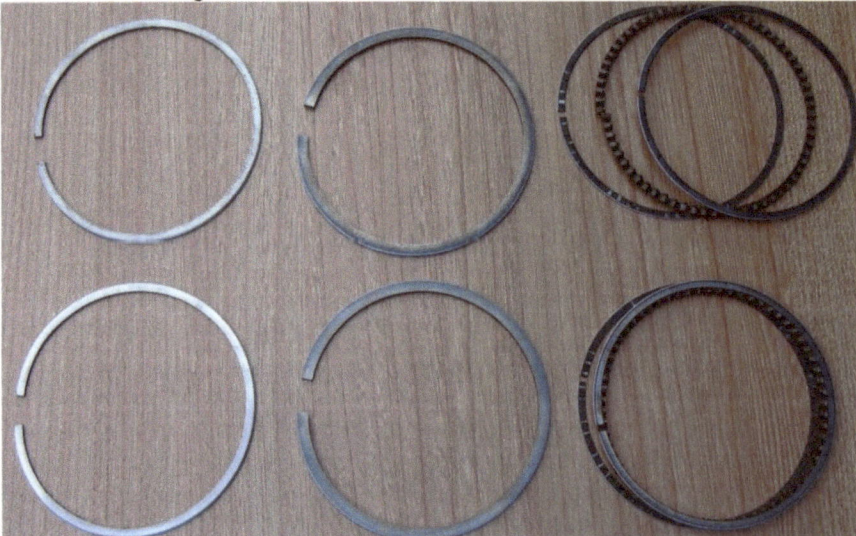

Piston rings
Figure 2-11

 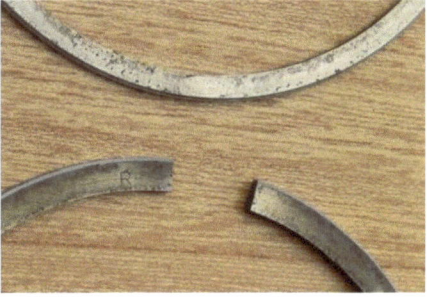

piston ring grooves
Figure 2-12

Marking of the piston rings

The numerical combination "25" shows the first oversize dimension of 75.25 mm. Figure 2-12 shows the piston ring grooves and the piston rings in detail. The upper ring is the flatter and narrower ring, which belongs to the upper piston ring groove. Further information about the pistons can be found in the original workshop manual on page 26.

2.1.2 The Cylinders

The cylinders consist of an aluminum housing into which gray cast iron cylinder liners are pressed. On the right-hand side of figure 2-13, a piston ring is inserted into a cylinder liner. The gap indicated by a red circle must not exceed a dimension described in more detail in the original workshop manual. If the gap is too large, the pistons and cylinder liners must be measured, as described in the original workshop manual, and reworked in case of excessive wear. The inner diameter of the cylinder liners is then increased to the next oversize dimension and oversize pistons are fitted with new, matching piston rings.

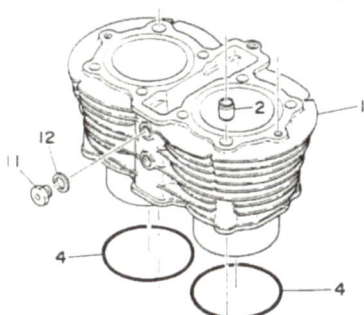 

Cylinder
Figure 2-13

2.1.3 Timing chain, camshaft

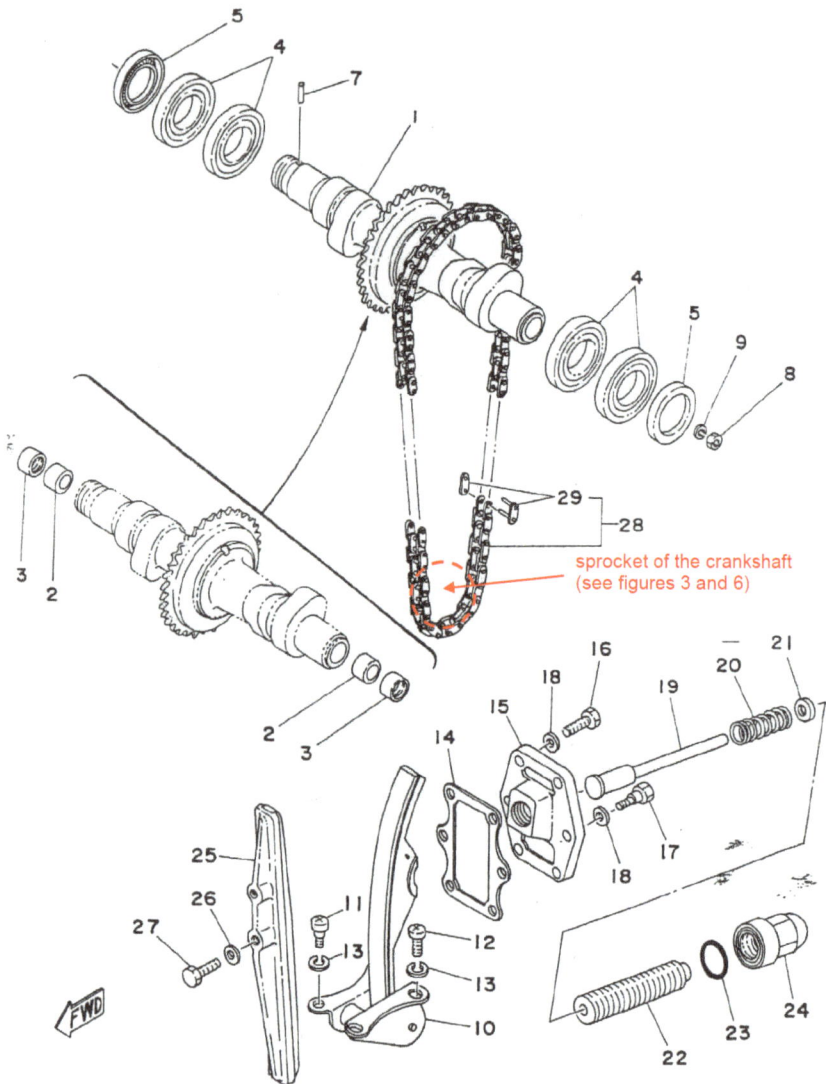

Camshaft and timing chain

Figure 2-14

Figure 2-14 shows the timing chain with its tensioning mechanism and the camshaft. The timing chain is an endless roller chain that has been riveted at the factory. A tensioning device (items 15 –24 above and Figure 2-15) is necessary since a certain pretensioning of the timing chain is necessary for operation, and since the chain gets longer with wear over time.

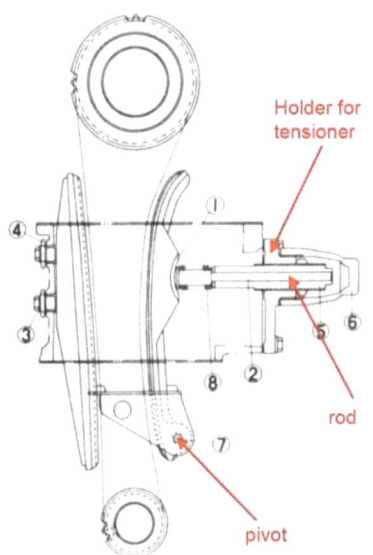 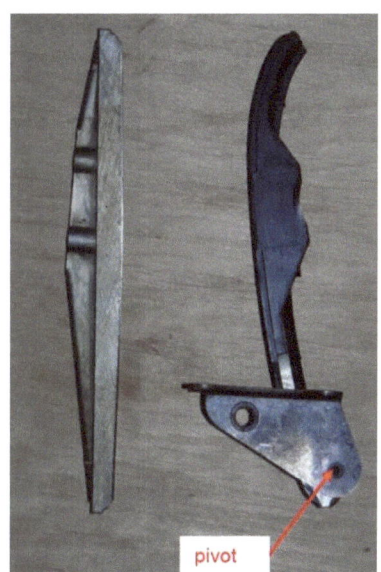

Timing chain tensioner
Figure 2-15

The timing chain is guided and preloaded by two guide rails. The guide rail at the front of the engine is bolted to the cylinder housing. The rear guide rail is rotatably mounted with its lower end in the upper part of the engine housing. The center of rotation is indicated by an arrow mark in Figure 2-15, which shows the guide rails in the form of a graphic and a photo. Both guide rails are made from aluminum, which is covered with a plastic sliding layer. The rotatably mounted guide rail has a shaping on the rear side, into which the tensioning mechanism engages. This tensioning mechanism consists of a housing part with an internal thread, which is connected to the cylinder housing via six M 6 bolts.

Tensioner of the timing chain
Figure 2-16

The actual tensioning element consists of a hollow-bored bolt with an external thread and a hexagon at one end (item 1 in figure 2-16) which is screwed into the housing with an internal thread. The threads serve to adjust the tensioning of the timing chain. In the tensioned state, the timing chain is loaded by a pressure spring (item 2), which is actuated by a pin (item 3) which is mounted in the hollow-bored bolt and has a mushroom-shaped configuration at its end. The "mushroom" engages in a corresponding shaping of the rear, rotatably mounted timing chain guide rail. Between the spring and the hollow-bored bolt there is a rubber-coated disc (item 4) which is intended to prevent the noise generated by the timing chain from being transmitted to the housing. Pretensioning of the timing chain is necessary since the crankshaft does not rotate at a constant speed, especially at low revs, which would lead to chain striking if the chain is not tensioned properly. However, an overtensioned timing chain leads to excessive wear of the timing chain guide rails and the timing chain itself.

According to the original workshop manual the pin (item 3) is intended to close flush with the hexagon of the bolt to indicate that the tension of the timing chain is correct. However, since the force of the spring (item 2) decreases with time, this can only be a reference value. It is better to tension the timing chain while the enginge is idling by turning the bolt while simultaneously touching the end of the pin (item 5) with a fingertip. If the pin still pulsates slightly, the timing chain is properly tensioned. You should rely on this feeling. There are different versions of the tensioning mechanism as far as the counter-bolt is concerned, but the mode of operation is always the same.

Camshaft with valve drive and valves
Figure 2-17

There are valve trains with a gear ratio of 17 to 34 teeth and those with a gear ratio of 18 to 36 teeth. The gear ratio 18 to 36 teeth was introduced together with the 1974 model year (TX650A). The chains belonging to the individual variants are also different so that swapping is not possible.

The overhead camshaft, which is driven by the timing chain, is mounted in four ball bearings in the cylinder head. The cylinder head consists of an upper part and a lower part. The valves and the lower half of the camshaft bearing seats are located in the lower part of the cylinder head. In the upper part there are the rocker arm shafts with the rocker arms and the upper half of the bearing seats of the camshaft bearings.

Figure 2-17 on page 14 shows the timing chain on the sprocket of the camshaft in the center of the engine. To the right and to the left, the cams of the camshaft with the ends of the rocker arms can be seen in the center of the image. Further down you can see the rear rocker arm shafts. At the bottom of the image, the valve disks and valve springs and the other ends of the rocker arms can be seen with the valve clearance adjustment screws. In the upper part of the image you can see the connections of the oil pipe to supply oil to the valve train.

The camshaft is designed as a forged hollow shaft, which is cylindrically shaped at both ends for receiving the 4 camshaft bearings, which are ball bearings. The cams for actuating the inlet and outlet valves of the right and left cylinders are located further inside. The sprocket of the timing chain is shrunk onto the shaft in the middle. On the right side there is a thread for receiving the centrifugal governor of the ignition, which is connected to the base plate of the ignition contacts in the housing on the left side of the cylinder head by a shaft inside the hollow-bored camshaft.

Camshaft
Figure 2-18

2.1.4 Valve actuation

The valves are arranged at an angle of approximately 80° in the cylinder head. The larger inlet valve has a diameter of 41 mm, while the smaller outlet valve has a diameter of 36 mm. Two pressure springs, designed as coil springs, which are provided for each of the valves, are installed with mutually opposing coils. Below the valve springs there is a steel disc, which prevents the valve springs from damaging the aluminum material of the cylinder head. At the upper end, the valves are held by a valve plate with conventional valve wedges. Figure 2-19 shows the valves together with the rocker arms and the rocker arm shafts.

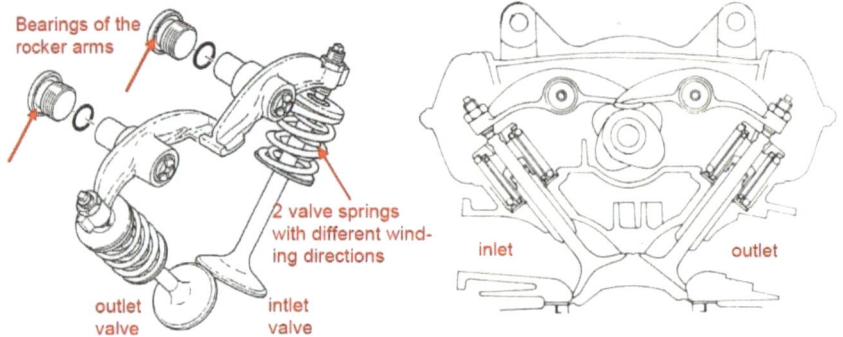

Figure 2-19

Figure 2-20 shows the valves of different sizes - the smaller outlet valve at the top and the larger inlet valve at the bottom. To the right there is a mounted valve with the valve plate and the valve wedges.

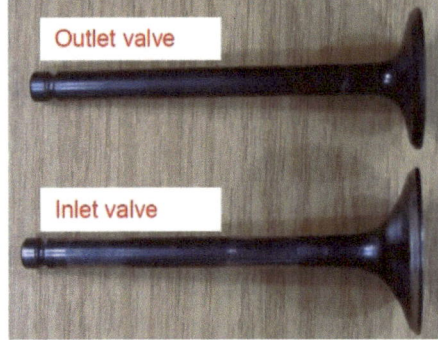 

Valves
Figure 2-20

Valve spring retainer with locks

Figure 2-21 shows a rocker arm with a valve adjusting screw in a side view and from above. There are different types of adjusting screws with square or hexagon sockets.

Rocker arm with valve adjusting srcew
Figure 2-21

The views in the upper and lower parts of the cylinder head are shown in Figure 2-22. The lower part of the cylinder head with the camshaft and the valves still mounted can be seen on the left, the timing chain being already open. To the right there is the view into the top of the cylinder head with the bearing seats of the camshaft bearings and the rocker arms.

View on the lower part of the cylinder head   **View on the upper part of the cylinder head**
Figure 2-22

By using a sectional model of the front left valve cover, figure 2-23 shows the upper part of a valve (valve disc) and the end of a rocker arm with the adjusting screw for setting the valve clearance (green arrow).

Left outlet valve with cutaway cover
Figure 2-23

2.2 Power Transmission

The rotation of the crankshaft is transmitted to the rear wheel via the primary drive, the clutch, the transmission (gearbox) and the drive chain. The components up to the sprocket of the drive chain are described below.

2.2.1 Primary drive

The primary drive (Figure 2-24, yellow arrows) consists of straight toothed spur gears of which the smaller one with 27 teeth is fixed to the right side of the crankshaft by a wedge. The larger one with 72 teeth is attached to the rear of the clutch assembly. The torque is transmitted from the smaller gearwheel of the primary drive to the larger gearwheel and to the pressure plates of the clutch hub via 6 compression springs arranged in the circumferential direction on the backside of the clutch assembly.

Primary drive
Figure 2-24

2.2.2 The Clutch

The clutch consists of an outer part, the clutch housing, which is connected to the large spur gear of the primary drive in a torsion-resistant manner by means of six damping compression springs arranged in the circumferential direction. On the transmission input shaft, the clutch boss is rotatably supported by means of a brass bushing.

Via the inner toothing of the inner part, the inner part of the clutch is non-rotatably connected to the transmission input shaft, which is provided with an outer toothing. In the outer part, the clutch housing, the inner part (the clutch boss) is rotatably mounted with a radial bearing.

The blue arrow in Figure 2-25 shows the toothing of the transmission input shaft and the green arrow shows the outer thrust washer of the radial bearing.

The clutch
Figure 2-25

There are clutches with 6 and 7 friction plates, which are the same concerning the function and the structure however. The friction plates of the clutch engage with the external pins on the circumference in corresponding gaps of the clutch housing (yellow arrow in figure 2-25). Between the friction plates, pressure plates with an internal toothing are arranged, which engage in an external toothing of the clutch boss (turquoise arrows in figure 2-25).

The torque connection is established by pressing the friction plates and the pressure plates together by means of the clutch pressure springs (orange arrows in figure 2-25). The individual parts of the clutch are reproduced by means of an explosion drawing from the spare part list in figure 2-26.

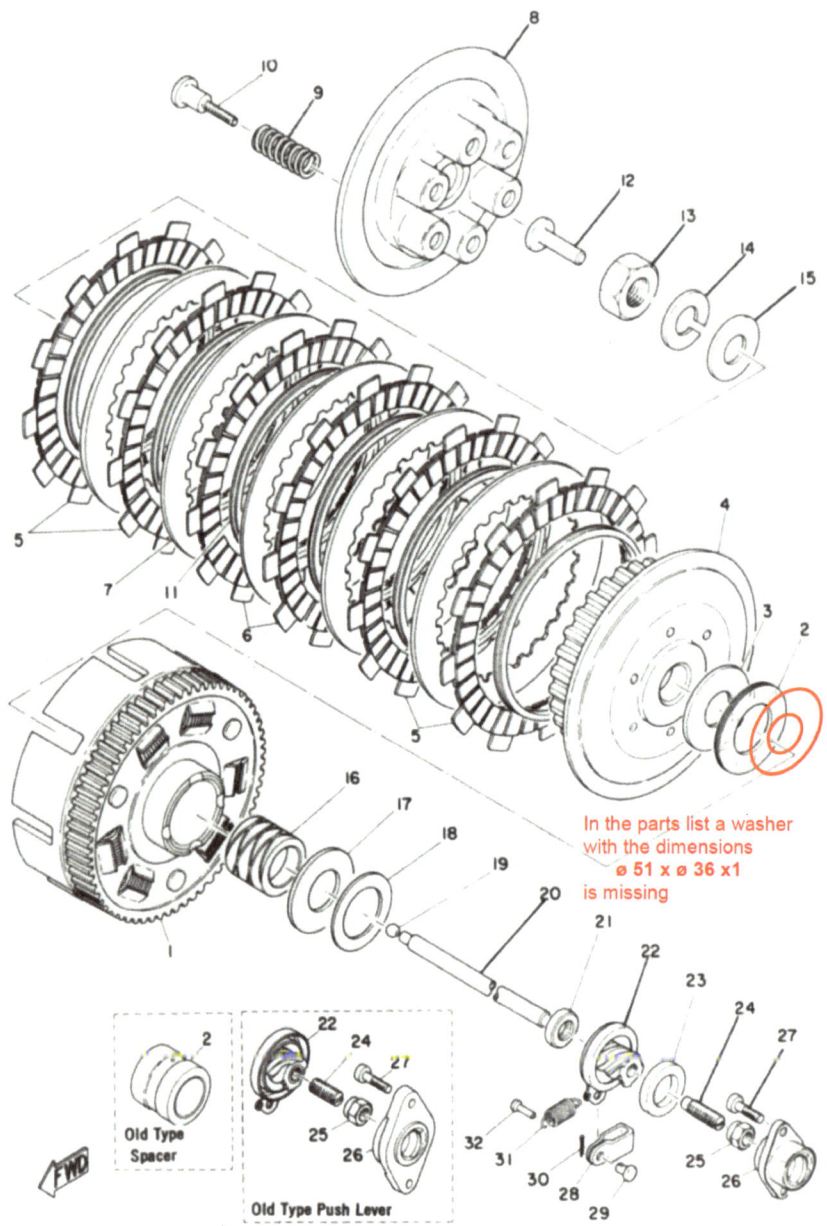

Figure 2-26

Figure 2-27 shows the back of the clutch housing with the damping springs arranged in the circumferential direction and the brass bushing with which the clutch housing is rotatably mounted on the transmission input shaft. The sheet metal disc with in the circumferential direction arranged recesses for the damping springs is fixedly connected by three rivets. The gear of the primary drive is rotatably arranged between the sheet metal disc and the outer clutch housing.

Figure 2-27

When looking at the end face of a spring, the upper third of the damping spring is located in the recess of the sheet metal disc shown in figure 2-27. The middle third of the damping spring is surrounded by a rectangular cutout in the end face of the gearwheel of the primary drive. The lower third of the damping spring is then again located in a corresponding recess of the clutch housing.

Figure 2-28 on the next page shows the funktion of the clutch by means of a sectional drawing. The end face of a damping spring and its arrangement as described above is indicated by a red color marking. The clutch housing, indicated by blue color marking, is driven by the primary drive. The clutch plates, also marked in blue, rotate together with the clutch housing. The inner part of the clutch (marked yellow) is firmly connected to the orange-colored transmission input shaft.

The clutch push rod is located inside the transmission main axle, which is hollow. The clutch is disengaged by moving the clutch push rod (light green color mark) from left to right against the yellow-marked plate of the clutch.

The clutch pressure springs are compressed and thus the pressure between the friction plates and the pressure plates is released, so that the clutch housing can rotate against the clutch boss and the transmission main axle.

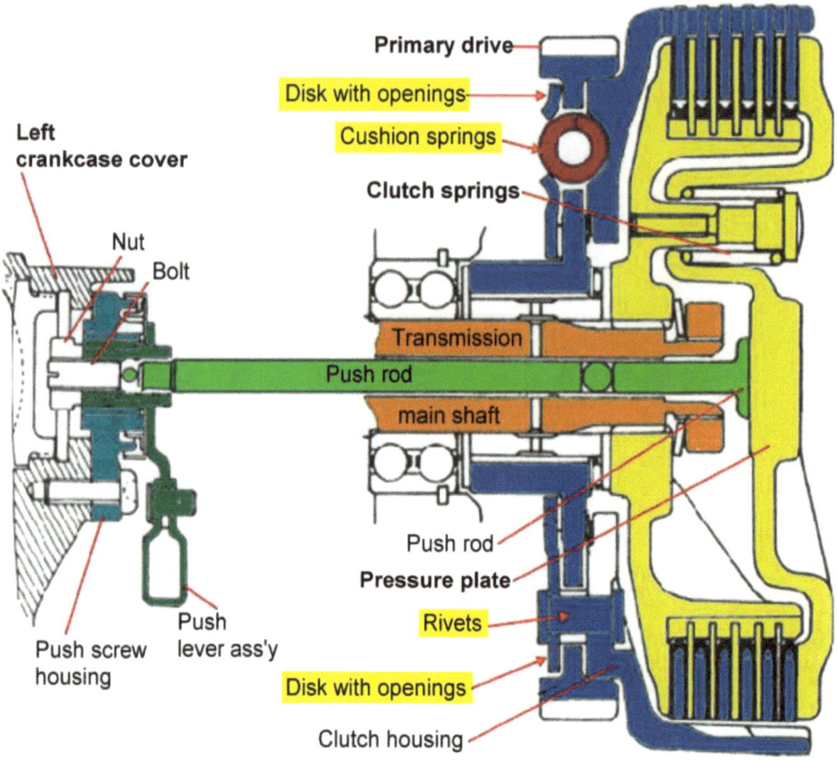

Function of the clutch
Figure 2-28

2.2.3 Clutch actuation

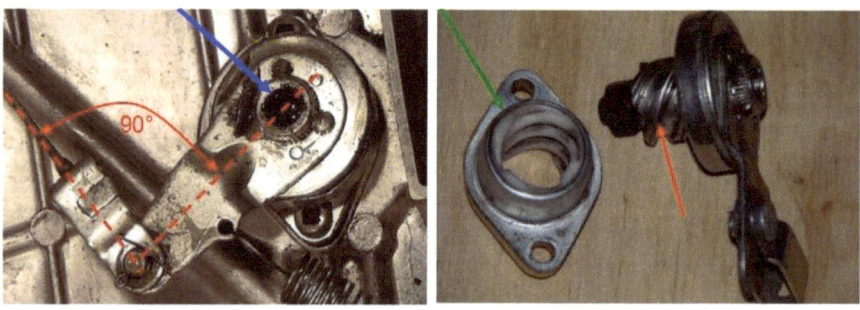

Release mechanism of the clutch
Figure 2-29

Some owners replace the two-piece clutch pressure rod with a one-piece rod so that one can find both variants when working on a second-hand engine.

If the clutch does'nt release properly, the reason is most often too much friction inside the bowden cable or an incorrect adjustment of the angle between the bowden cable and the lever of the disengagement mechanism. **Since the force for disengaging the clutch increases with increasing distance, the transmission between the hand lever and the disengagement mechanism must get "more favorable" with the increasing travel of the hand lever, and reach its maximum with the hand lever fully pulled.** This is achieved when the angle between the bowden cable and the lever of the disengagement mechanism is 90° when the lever is fully pulled.

2.2.4 Gear box

Figure 2-30 shows the five-speed gearbox together with the shifting mechanism, the primary drive, the clutch and the kickstarter in the built-in state using a cutaway model.

Figure 2-30

2.2.5 Transmission ratios

The following table shows the gear ratios in the individual gears.

1. speed:	2.462
2. speed:	1.588
3. speed:	1.300
4. speed:	1.095
5. speed:	0.957

The numbers for the gears have different characteristic colors, which are continuously used in the following description of the transmission. The ratio of the first gear is 1: 2.462. That means, that the motorcycle reaches a speed of about 66 km/h at the rated speed of the engine (7000 rpm) with the standard secondary drive (ratio of 17 to 33) and the standard tires (Figure 2-31).

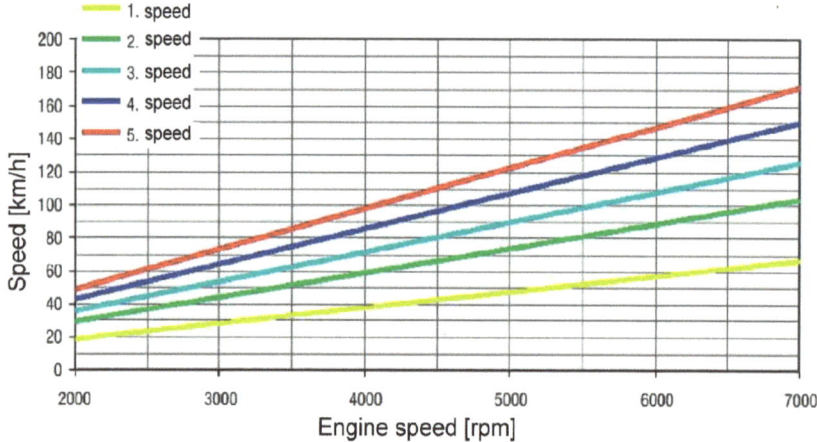

Speed chart
Figure 2-31

The distance to the second gear (1: 1.588), which reaches more than 100 km/h at the rated speed of the engine, is comparatively large compared to the distance between the second and third gears as well as between the other gears. Such a gradation is rather unfavorable for everyday operation in today's urban traffic. The reason for such a layout of the transmission was that in the early 1970s it was important to be able to specify a time as short as possible to reach a speed of, for example, 100 km/h.

Figure 2-32 shows for comparison purposes, the transmission ratio diagrams of older German motorcycles, which were still used in day-to-day operations for the daily way to work and less as a leisure vehicle with sporting ambitions. The left diagram shows the gradation of the gear ratios of a BMW R 25/3 from the 50s. Here, the distance between the individual gears of the four-speed transmission is approximately constant, which is particularly advantageous when driving in urban traffic. In the right-hand gear diagram shown in Figure 2-32 (Maico MD 250), the gradation of the gears becomes narrower, as in the transmission of the XS 650, which is more advantageous for sporty driving in higher speed ranges.

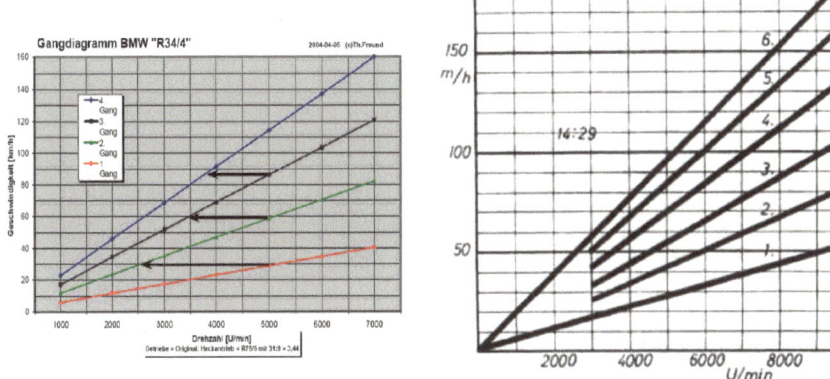

Comparison: Speed charts of older german motorcycles
Figure 2-32

2.2.6 Structure

Figure 2-33 shows the schematic structure of the five-speed transmission. Of the gear pairs of the individual gears, one gearwheel is rotatably mounted on the transmission input shaft or the transmission output shaft so that it can rotate freely on the shaft. The torque connection is established by shifting the gearwheel of another gear in the longitudinal direction of the shaft. The cams provided on the end face of one gearwheel can then engage with recesses of another gearwheel on the same shaft.

The four gearwheels of the first and second gears positioned at the ends of the transmission input- and output shafts cannot be moved in the axial direction.

On the transmission input shaft, two gears, the gearwheel for the 4th gear (blue marking) and the gearwheel for the 5th gear (red marking) are rotatably mounted. On the transmission output shaft there are three gearwheels rotatably mounted: the gearwheel for the 2nd gear (green marking), for the 3rd gear (turquoise marking) and for the 1st gear (yellow marking).

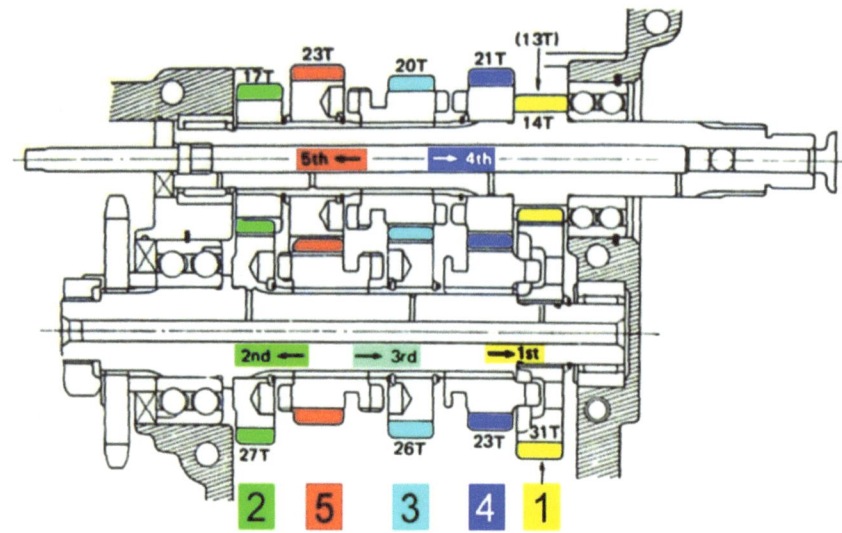

**Schematic figure: speeds and belonging gear pairs, shifting operations
Figure 2-33**

2.2.7 Funktion, drive train

In principle, the power train is established in such a way that one of the gearswheels, which is rotatably mounted on one of the two shafts of the transmission is positively engaged with a gearwheel of a different gear, which is moveable but non-rotatably mounted on the same shaft. The gears are shifted as shown in figure 2-34 on the next page.

1st gear:
The gearwheel of the 4th speed on the transmission output shaft is moved in the direction of the yellow arrow on the figure 34 and engages with its cams in recesses of the gearwheel for the 1st gear.

2nd gear:
The gearwheel of the 5th speed on the transmission output shaft is moved in the direction of the green arrow on the figure 34 and engages with its cams in recesses of the gearwheel for the 2nd gear.

3rd gear:
The gearwheel of the speed on the transmission output shaft is shifted in the direction of the turquoise arrow on the figure 34 and engages with its cams in recesses of the gear wheel for the 3rd gear.

4th gear:
The gearwheel of the 3rd gear on the transmission input shaft is shifted in

the direction of the blue arrow on the figure 34 and engages with its cams in recesses of the gear wheel for the 3rd gear.

5th gear:
The gearwheel of the 3rd gear on the transmission input shaft is shifted in the direction of the red arrow on the figure and engages with its cams in recesses of the gear wheel for the 5th gear.

Speeds and belonging gear pairs, shifting operations
Figure 2-34

For shifting the gears, a total of three shifting forks are required. Two of the shift forks engage in the circumferential grooves of the fourth and fifth speed gearswheels on the transmission input shaft, and a third shift fork engages in a circumferential groove of the third speed gearwheel on the transmission output shaft .

Figures 2-35 and 2-36 on the next page show the gearshift forks in a position on the transmission shafts (figure 2-35) and in the built-in position in the upper half of the engine housing (figure 2-36).

The shifting forks move the gearwheels in axial direction of the respective shafts and establish the torque connection by the engaging cams in corresponding recesses of the gearwheel of the gear to be shifted. The arrows on the shiftforks indicate the direction of the movement of the shiftforks and the marking colors of the arrows indicate the individual gears.

Speeds and belonging shift forks, shifting operations
Figure 2-35

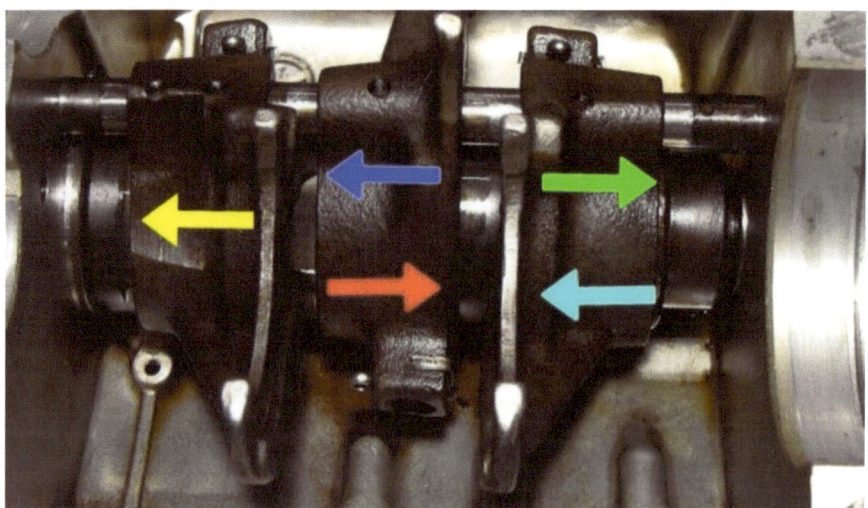

Shift forks and shifting operations
Figure 2-36

Figure 2-37 on the next page shows the gearwheel of the 1st gear with recesses and the gearwheel of the 4th gear with cams on the transmission output shaft in the idle position (left) and in the engaged state of the 1st gear with the cam engaged.

Notches of the gear of the 1. speed and switching cams of the gear of the 4. speed

1. speed engaged

Figure 2-37

2.2.8 Shifting mechanism

Figure 2-38 on the next page shows the components of the shifting mechanism of the shiftcam - in an exploded drawing from the spare parts list.

The shifting mechanism serves to convert the up-and-down movement of the foot shift lever into sliding movements of the shifting forks. The shifting forks have a large and a small bore at their upper end. The small bore slides on a tube mounted on both sides in the engine housing (item 14 in figure 2-38). The actual shiftcam (item 1 in figure 2-38) is guided within the large bore.

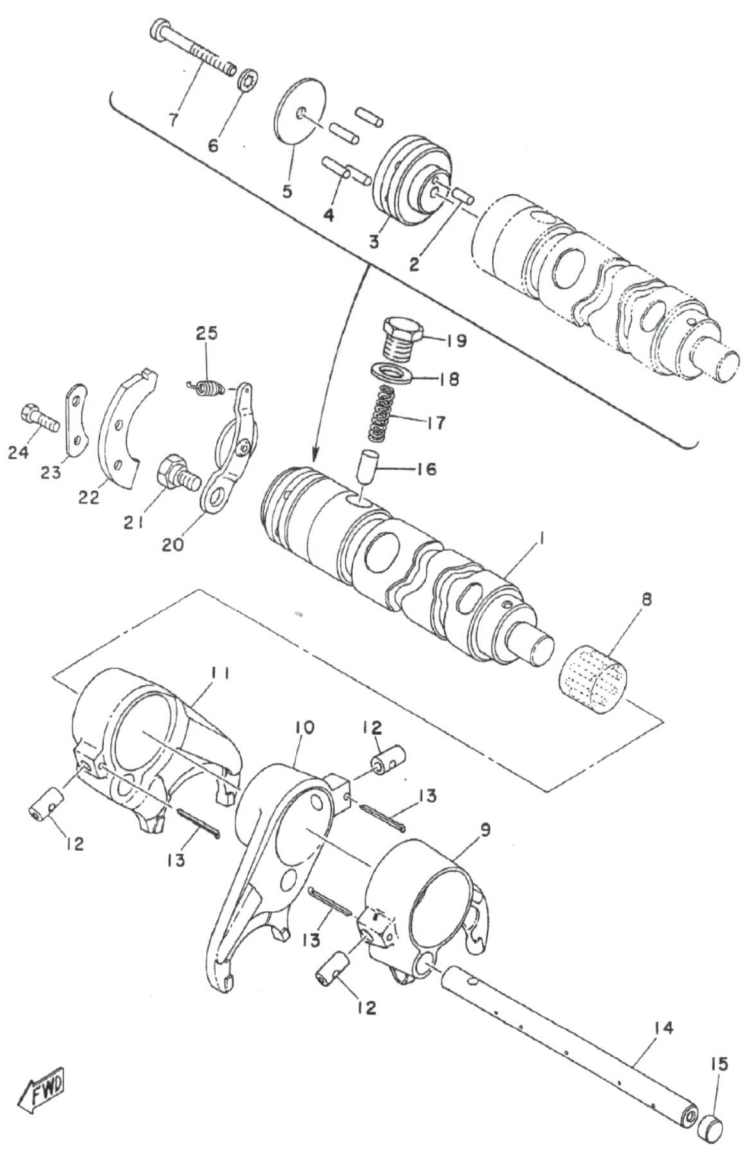

Shift cam assembly
Figure 2-38

On the shiftcam there are circumferential guide grooves, which arcuate at certain points on the circumference of the shiftcam (arrow markings in figure 2-39).

Figure 2-39

In the guide grooves of the shift cam, guide pins (item 12 in figure 2-38) are engaged with one end, while the other end is engaged in bores of the shift forks as shown in figure 2-40.

Figure 2-40

A shift fork is moved in the axial direction and a gear is thereby sfifted when a guide pin is engaged within the arcuate course of the guide grooves.

The shift shaft serves to transmit the up-and-down movement of the foot shift lever to the shift cam. By actuating the foot shift lever, the arm welded to the shift shaft is pivoted so that the fork attached to the upper end of the arm is moved for upshifting and downshifting as shown in figures 2-41 and 2-42. The arm, which is welded to the shift shaft, is held in the middle position by a spring, so that the same fork travel is available for upshifting and downshifting.

Figure 2-41

Figure 2-42

The adjusting screw with an eccentric shank indicated by an arrow in figure 2-43 is used to adjust the center position of the arm welded to the shift shaft.

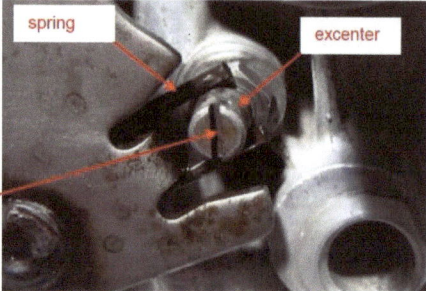

Adjusting the shifting mecanism
Figure 2-43

Neutral position of the transmission

On the left side of the upper engine housing half there is an electrical switch, which is closed by the pressure of a pin on the shift cam. On the right hand side of the upper engine housing half there is a spring loaded bolt, which engages in a recess of the shift cam when the transmission is in the neutral position. Figure 2-44 shows the shift cam in the neutral position with the neutral switch and the spring-loaded bolt engaged.

Shift cam in neutral
Figure 2-44

Figure 2-45 shows the neutral switch and the spring-loaded bolt with the shift cam removed.

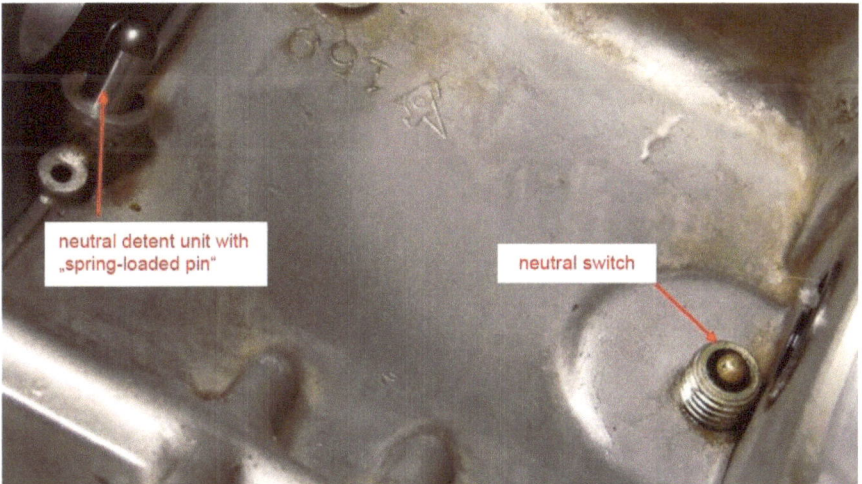

View into the upper crankcase
Figure 2-45

Figure 2-46 shows the actuation of the neutal switch on the shift cam and the recess into which the spring-loaded bolt engages when the transmission is in the neutal.

Figure 2-46

Behind the "star" on the right endface of the shift cam, there are cylindrical pins (arrow in figure 2-47) between which a disc loaded by a spring snaps in place when a gear is engaged, as shown on the right. If the transmission is in neutral, the disc is placed on a pin as shown in the picture on the left.

Figure 2-47

The pin on which the disc is located when the transmission is in neutral is not round, but half-round, with the disc touching the flat side of the pin. When shifting from one gear to the next, the disc "rolls" over the round pins. When idling, the disc "stands" on the flattened side of the half-round pin.

If it is difficult to get the transmission in neutral, that is, if the neutral position is skipped when downshifting from the 2nd gear or upshifting from the 1st gear during standstill, this may be caused by a too low a preload of the spring loaded bolt. If the pretension of the spring, which holds the discs in the engaged position between the pins, is too low, there is a risk that the gears will snap out, especially when the gearshift fork is already worn.

If the neutral position still engages properly when the engine is cold, but not when the engine gets hot, the reason may be, that the oil has got too thin due to the heat and there is no longer sufficient lubrication. This effect is particularly unpleasant in urban traffic, where the engine can become very hot because of the lack of an air stream.

2.2.9 Kickstarter

Figure 2-48 shows the components of the kickstarter mechanism using an exploded view from the spare part list.

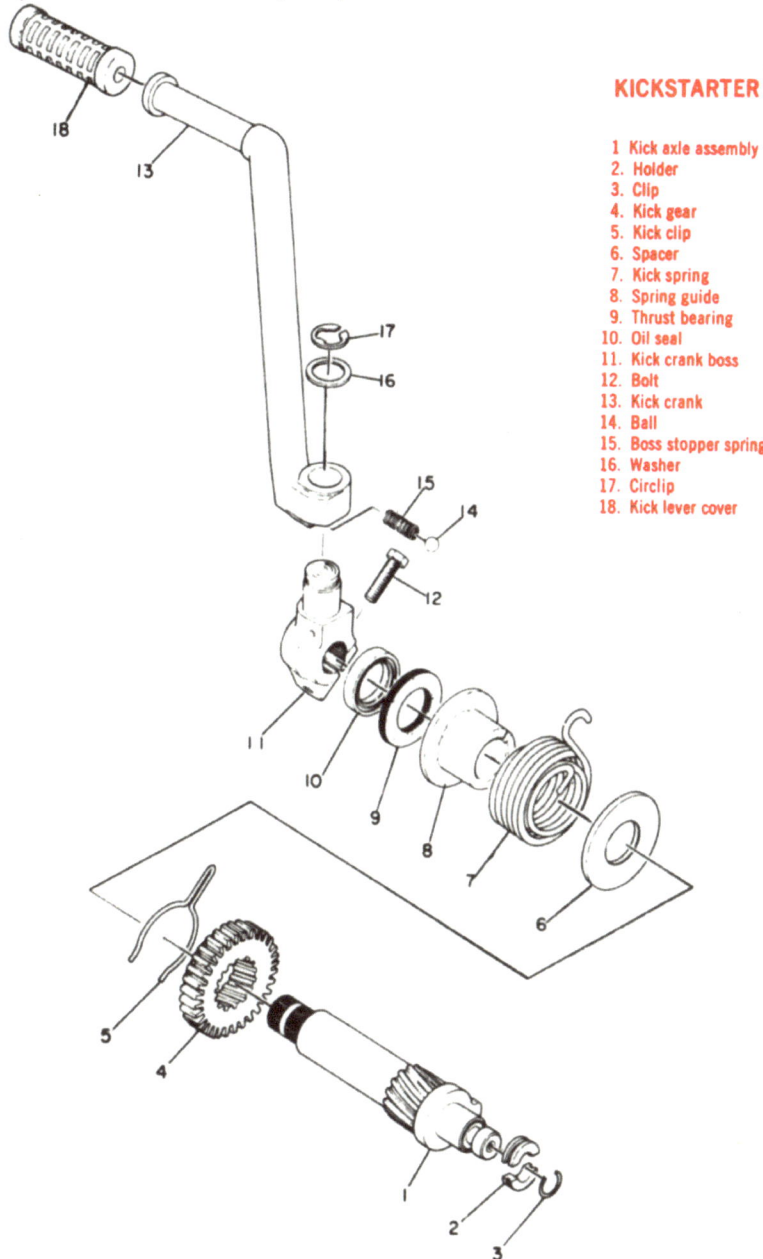

KICKSTARTER

1. Kick axle assembly
2. Holder
3. Clip
4. Kick gear
5. Kick clip
6. Spacer
7. Kick spring
8. Spring guide
9. Thrust bearing
10. Oil seal
11. Kick crank boss
12. Bolt
13. Kick crank
14. Ball
15. Boss stopper spring
16. Washer
17. Circlip
18. Kick lever cover

Figure 2-48

The kickstarter mechanism consists of the kickstarter shaft, which is mounted in the lower part of the engine housing and in the left engine side cover. Wrapped around the shaft is a coiled spring (item 7 in figure 2-48), which returns the kickstarter lever to its starting position. Attached to the kickstarter shaft (item 1 in figure 2-48) there is a gear wheel, which is connected to the kickstarter shaft by a helical toothing. If the kickstarter shaft is rotated, the gear wheel is moved in the axial direction on the kickstarter shaft.

The gear wheel is prevented from rotating together with the kickstarter shaft by means of a clip (item 5 in figure 2-48). Thus it moves on the helical gearing of the shaft in the direction of the gear wheel of the first gear on the transmission main axle (figure 2-49).

Figure 2-49

Figure 2-50 shows the kickstarter mechanism in the built-in state on a cutaway model, as well as the clip (yellow arrow marker and item 5 in figure 2-48)

Figure 2-50

2.3 Oil circuit

The XS 650 engine is equipped with a pressure lubrication system with an oil pump in the right engine side cover, which is driven by the crankshaft. The engine and the transmission have a common oil circuit.

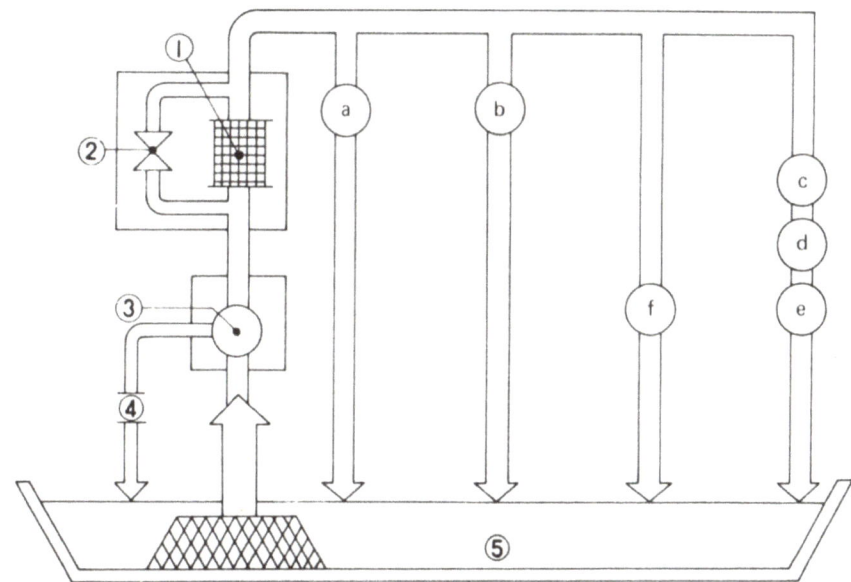

Oil Lubrication System
Figure 2-51

The oil pump (item 3 in figure 2-51) draws the engine oil through a filter located at the lowest point of the engine housing under the crankshaft. In the direction of the flow behind the oil pump there is a further sieve filter (item 1 in figure 2-51) which is equipped with a bypass (item 2).

Behind the sieve filter the engine oil is supplied via oil ducts into the upper part of the engine housing and separate pipes to the individual lubrication points, which are described in detail in the following. Since no hydrodynamic bearings exist, no pressure builds up in the oil circuit, which would be measurable with a commercially available oil pressure gauge.

The cause of many engine and transmission damages can be oil deficiency due to clogged oil ducts. Before exchanging worn parts, you should definitely check all oil ducts and clean them if necessary. The oil circuit is therefore described in detail in the following.

2.3.1 The oilfilter

Figure 2-52 shows the location of the oil filter at the lowest point of the engine housing below the crankshaft and with a typical damage - a torn screen - in the bolted state on the base plate.

Oil filter in the lower crankcase

Oil filter bolted to the base plate

Figure 2-52

The original oil filter is often replaced by paper filters, which are supplied by dealers who are specializing in the XS650. These filters have the advantage that they filter the oil much finer than the original sieve filter.

The disadvantage is that a completely filled paper filter simply prevents the oil pump from drawing oil if it is located on the suction side of the oil pump. A filter arranged behind the oil pump will tear when it is full and let the oil pass.

Behind the oil filter, the oil flows through a channel in the base plate into an opening on the lower side of the lower half of the engine housing (arrows in figure 2-53).

Base plate with oil line

Oil inlet in the crankcase

Figure 2-53

The oil pump, which is located in the right engine side cover, draws the oil through a channel as shown in figure 2-54 on the next page.

From the outlet opening of the lower engine housing half, the oil flows into an inlet opening in the right engine side cover.

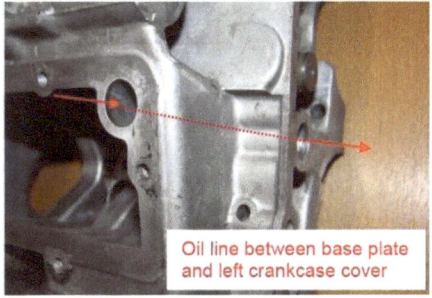

Figure 2-54

Figure 2-55 shows the oil ducts and the oil pump in the right engine side cover.

Figure 2-55

2.3.2 The oil pump

The components of the oil pump, consisting of the housing, the outer and the inner rotor, as well as the drive shaft that drives the inner rotor are shown in figure 2-56.

Figure 2-56

Figures 2-57 to 2-60 show the oil pump in the built-in state by means of a cutaway model and in the assembled state viewed from the inside. Figure 2-58 shows the drive shaft with a wedge for the connection to the drive gear wheel of the oil pump.

Figure 2-57

Figure 2-58

There are arrows on the inner and the outer rotor, which must be brought into alignment during assembly. The arrows in Figure 2-58 show the path of the oil through the oil pump and the second oil filter.

The oil pump is driven by the gear wheel shown in figure 2-59, which engages with a further gear wheel on the crankshaft. Together with the drive gear, there is a worm gear on the drive shaft of the oil pump that drives the speedometer (figure 2-60).

Figure 2-59

Figure 2-61 shows the secondary oil filter behind the oil pump, which is fixed in the right engine side cover by means of a hollow pin in which there is a bypass valve. From the secondary oil filter, the oil flows into the upper engine housing half, where it is distributed to the individual lubrication points.

Figure 2-60

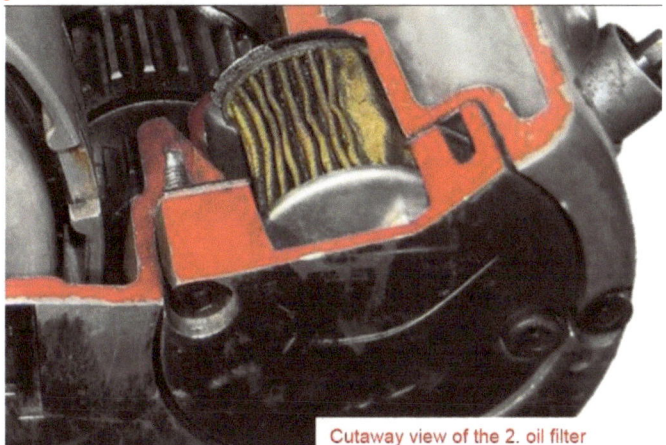

Figure 2-61

2.3.3 Lubrication points of the engine

From the 2. oil filter in the right crankcase cover (figure 2-62) the engine oil flows into an oil passage extending transversely in front of the engine housing (figure 2-65). From there the two central crankshaft bearings and oil nozzles which are lubricating the connecting rod bearings are supplied. In the middle of the horizontally extending oil channel, the oil rising pipe branches off in front of the cylinders to supply the lubrication points in the cylinder head. The right crankshaft bearing (fixed bearing) is lubricated by the oil mist in the right engine side cover. The right crankcase cover, the removed 2. oil filter and the bypass valve are shown in figures 2-62 and 2-63. Figure 2-64 shows the outlet opening of the engine oil in the right crankcase cover and the inlet opening into the upper engine housing half.

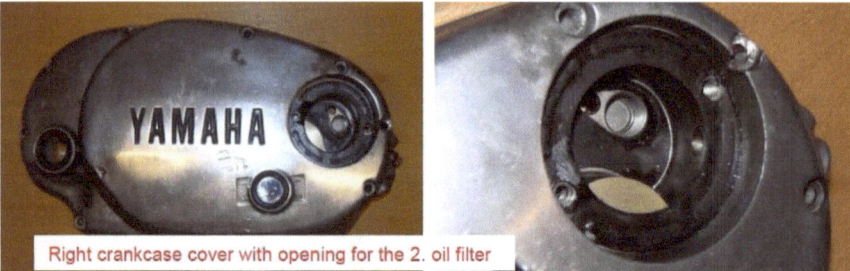

Right crankcase cover with opening for the 2. oil filter
Figure 2-62

2. oilfilter
Figure 2-63

Bypass valve

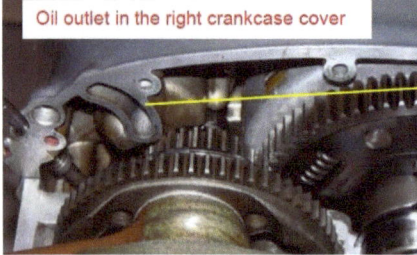

Oil outlet in the right crankcase cover

Figure 2-64

Oil inlet into the crankcase

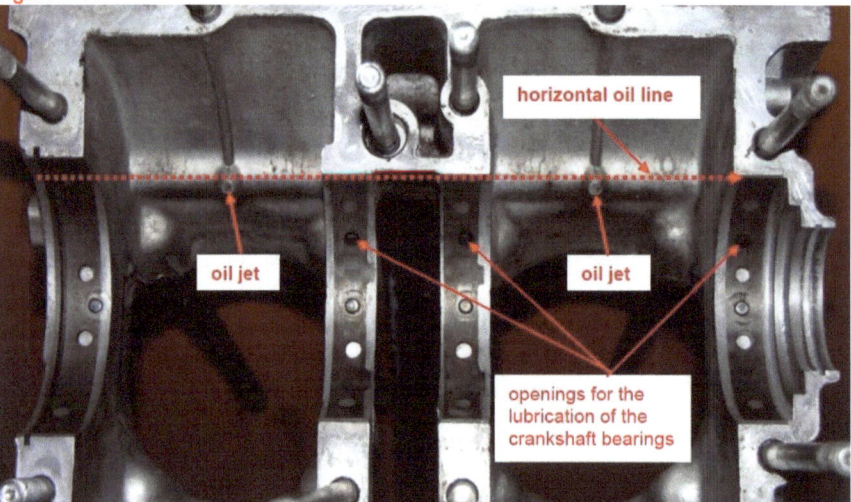

horizontal oil line

oil jet

oil jet

openings for the lubrication of the crankshaft bearings

Figure 2-65

Crankshaft, conrod bearings

The oil nozzles and the connecting rod bearings lubricated by these are shown in figures 2-66 and 2-67.

Figure 2-66

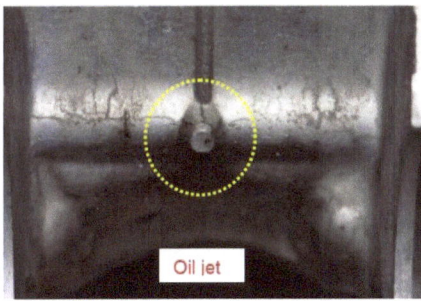

Figure 2-67

Rocker levers, valves

The oil rising pipe in the middle of the oil passage, which extends horizontally in front of the engine housing is the connection to the oil supply to the cylinder head, as shown in figure 2-68.

Junction of the oil pipe to the cylinder head   Connection of the oil pipe and cylinder head
Figure 2-68

Figure 2-69 shows the path of the engine oil to the lubrication points of the rocker arms and from there to the cams of the camshaft. The four camshaft bearings (grooved ball bearings) and the valve stems are lubricated by the oil mist in the cylinder head.

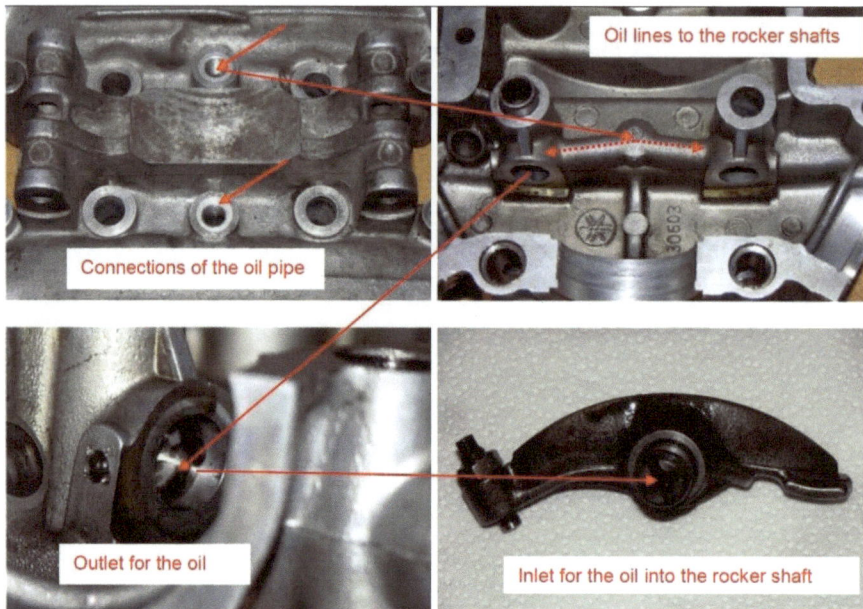

Figure 2-69

From the housing of the cylinder head, the engine oil enters the end face of the rocker arm shafts and then flows into the rocker arms through a groove (figure 2-70).

Figure 2-70

The rocker arms have an outlet opening in the vicinity of the sliding surfaces through which the sliding surfaces are lubricated.

Figure 2-71 shows the drain holes for the engine oil to flow from the cylinder head in the direction of the timing chain shaft and the vent holes that connect the volume of the motor housing to the environment.

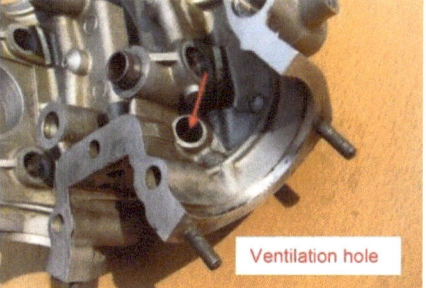

Figure 2-71

2.3.4 The lubrication of the transmission

From the oil passage extending horizontally in front of the engine housing, an oil passage branches off from the left crankshaft bearing and toward the needle bearing of the transmission main shaft on the left side of the upper engine housing half (figures. 2-72 and 2-73).

Figure 2-72

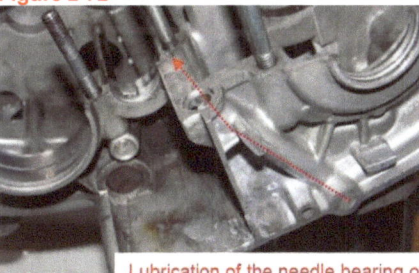

Figure 2-73

The bearings of the transmission shafts (ball bearings) are lubricated by the oil mist in the engine housing.

From the needle bearing of the transmission main shaft, an oil passage (figure 2-74) branches off to the guide tube of the shift forks. The guide tube has holes on the underside (circular markings in figure 2-75) through which the gearshift forks and gear wheels are lubricated and the gear wheels are cooled.

Needle bearing of the main shaft
Figure 2-74

Oil line through the guiding bar of the shift forks

guiding bar for the shift forks with oil jets
oil jets
Figure 2-75

Detail - oil jet for the needle bearing of the drive shaft

3 Components

3.1 Images

The following illustrations show the components of the XS 650 engine, arranged by module - similar to the explosion drawings in the spare parts list. The illustrations are intended to help when looking for spare parts and also to identify the parts. Basically, there are several types of spare parts: Standard parts, "quasi standard parts", spare parts which are also used in other models of the vehicle manufacturer and spare parts specifically intended for the XS650, which do not occur anywhere else. Particularly in the last genre - the spare parts specifically intended for the XS 650 and which do not occur anywhere else - one should never make a "commercial" decision if there is the choice between reworking and the purchase of a used spare part. A worn out camshaft can be reworked by an engine reconditioning service by grinding. It may still be cheaper to buy a used camshaft, which is still ok an throw the worn out away, but this situation may change quite fast, since camshafts are no longer produced.

3.1.1 The engine

Engine housing, cylinder head with camshaft and the cylinders are shown in Fig. 3-1.

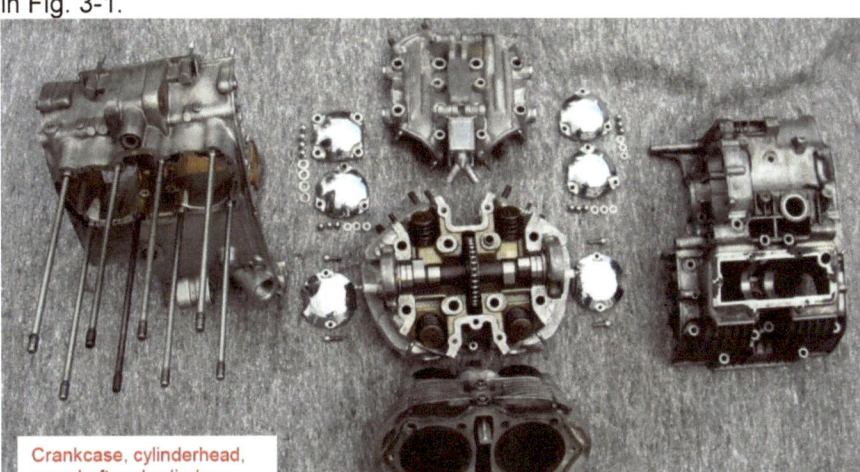
Crankcase, cylinderhead, camshaft and cylinders

Figure 3-1

The crankshaft with the pistons, including the bolts and retaining rings, as well as the camshaft with the bearings and the timing chain, can be seen in figures 3-2 to 3-4.

Crankshaft, pistons, timing chain and camshaft

Figure 3-2

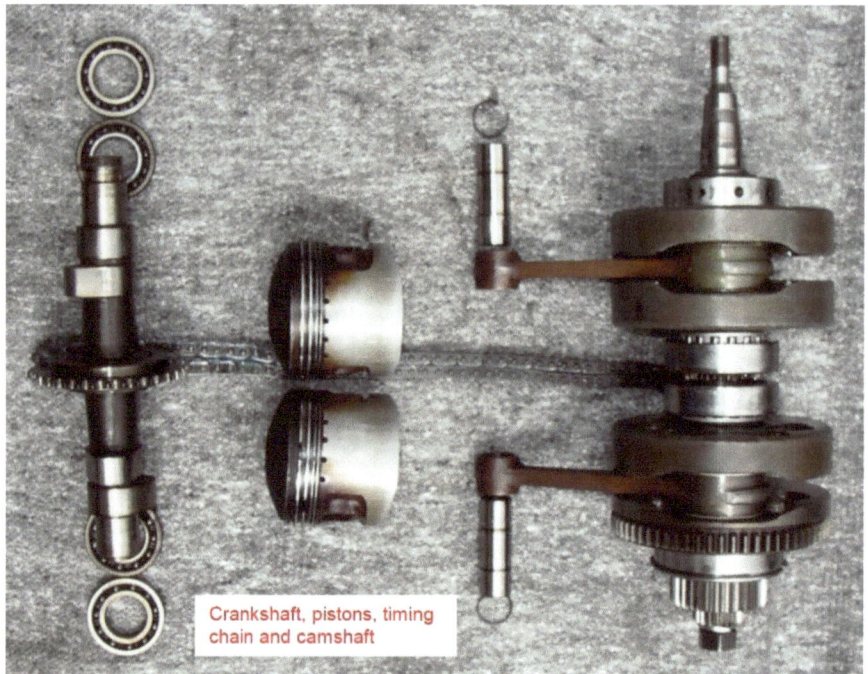

Crankshaft, pistons, timing chain and camshaft

Figure 3-3

Crankshaft, pistons, timing chain and camshaft

Figure 3-4

Figure 3-5 shows the crankshaft with the bearing removed.

Figure 3-5

The timung chain tensioner components are shown in figure 3-6.

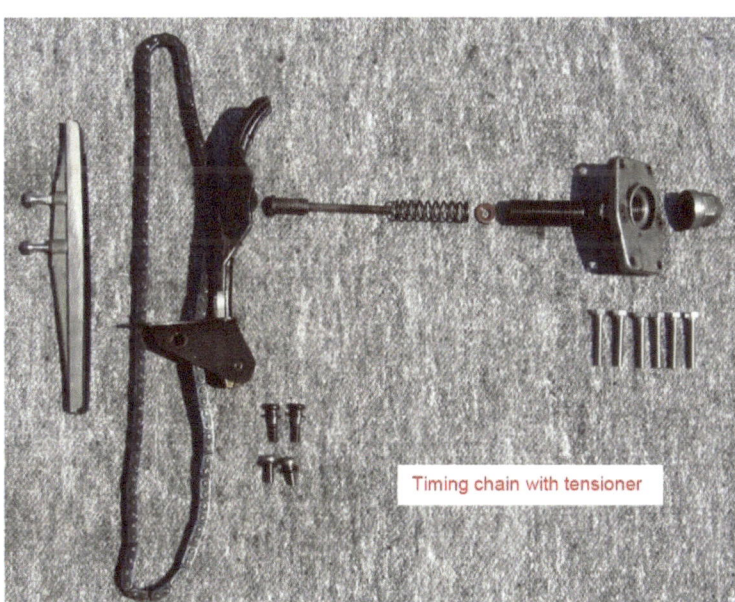

Figure 3-6

3.1.2 Transmission

All the components of the shifting mechanism are shown in figures 3-7 and 3-8.

Figure 3-7

Figure 3-8

Figure 3-9 shows the transmission in the assembled state.

Figure 3-9

3.2 Spare parts situation

The XS 650 was built in its latest version, the XS 650 SE Heritage, until 1984. Spare parts were supplied by Yamaha dealers, which also carried out repairs until the middle of the nineties. In the nineties spare parts were offered favorably by dealers and in the classified ads section of daily newspapers. It was possible to get a complete engine including the frame for two hundred Euro or even less.

The description of the current spare part situation of a vehicle type can always be only a snapshot, since it constantly changes. First, the spare parts supply is ensured by the manufacturer's dealer network. This is the case for up to 10 years after the end of the production. Then, the existing stock of spare parts is sold at a favorable price and some people buy large quantities.

In the XS 650 community there are mainly two groups of owners - those who look upon a XS 650 already as classic bike, which should be restored as original as possible. Then there are others, who want to create an individual motorcycle. Today there are still untouched originals, which are suitable as a restoration object. On the other hand, there are already more or less converted bikes, which are just kept driveable. Some others are converted as choppers, cafe-racers or otherwise without losing a basis for a restoration. Therefore, parts for conversions such as forward and backward mounted foot rests, tanks, etc. are offered by dealers specializing in the XS 650.

Today for example, it may still be cheaper to buy a used transmission or a used shift mechanism, as to rework a worn-out gearshift fork. Never one should throw away those parts, whose reprocessing is not yet worthwhile from a "commercial" point of view, if a repair is possible in principle. Thus, for example, a routine work in engine repair shops is to repair the cams of camshafts and sliding surfaces of rocker arms by grinding. A worn out gearshift fork can be repaired by adding material by welding and grinding it afterwards to get the proper shape.

Then there are "quasi-standard parts" such as pistons, piston rings and valves in standardized sizes offered by manufacturers specialized in such parts. These parts, which are available in an unlimited quantity can be obtained from the dealers mentioned below. It is not to be expected, that the price will increase significantly in the future.

Standard parts are e.g. all nuts and bolts, most shaft seals and some bearings. However, with the exception of the nuts and bolts and the paper gaskets, which can be cut from gasket paper, it is not worthwhile to buy standard parts such as bearings and shaft seals elsewhere than from spare parts dealers. Bearings and shaft seals generally are only sold through the wholesale trade in packaging units.

Another frequently used option for spare parts procurement are internet auction sites like Ebay, where spare parts are offered by professionel spare parts dealers as well as by private individuals. But you should not stick to the illusion that an XS 650 is a cheap old bike that you can keep as cheap as an old car. There are so far few reproduction parts, but these are produced in unit numbers which will hardly allow prices, which are lower than usual spare parts prices for new motorcycles. Reproduction parts made available by professional dealers can never be cheaper than spare parts supplied by the vehicle manufacturers for new motorcycles.

In general, the spare parts situation of the XS 650 models is still good with some restrictions. The restrictions are due to the fact that many want to use only original parts. In particular original exhaust systems, tanks and side covers are scarce and correspondingly expensive. The prices for engine and transmission parts have also risen in the recent years, as these are increasingly offered by dealers. This has the advantage, however, that nobody has to be afraid that he can not use his motorcycle during a season because a required spare part can not be procured. There are dealers like the ones mentioned in the chapter "3.3 Dealers", who have specialized in the sale of spare parts and accessories for the XS 650. In addition, these dealers are taking care of repairs and improvements, such as for example electronic ignitions.

At the moment, it looks as though the XS 650 is gaining in popularity. For example, carburettor diaphragms, transmission gear wheels, silencers, as well as various electronic ignitions are available. However, it is not foreseeable how the spare parts situation will develop in the future. It is conceivable that - as with many popular motorcycles of the 1950s - almost all spare parts will be again available if the Yamaha XS 650 develops to a classic.

3.3 Dealers

The following dealers offer spare parts and accessories for the XS 650:

Founded in 2006, TC Bros. has become the leading parts supplier for DIY, garage-built motorcycle enthusiasts all over the world.

Fast shipping and top notch customer service are what we are most known for. Get a custom, vintage look on your ride with USA-made Sportster, Big Twin, XS650 and CB750 parts.

Take a look through our online store for the high quality motorcycle parts that you need.

TC Bros. 12052 US Hwy 20A, Wauseon, Ohio 43567
419-265-9399 |
Sales@TCBrosChoppers.com   www.tcbroschoppers.com

27 years of workshop experience with Japanese motorcycles, and specializing in the Yamaha XS 650. Many special conversion projects, which several times have given the prices for the nicest remodeled and most beautiful original Yamaha XS 650.

Ringsted Motor started already in 1990 when the owner, Per Jacobsen, bought his first Yamaha XS 650, the "yellow lightning, 1976". Here the company logo comes from.

Ringsted Motor has extensive knowledge and experience with Yamaha XS 650. In addition, we hold the largest Nordic used storage of parts for Yamaha XS 650.

Ringsted Motor has 5 employees. Over the years 2-3 mechanics have made many repairs, troubleshooting, renovations, and a sea of engine repairs.

We can offer to solve all tasks related to Yamaha XS 650, and your Yamaha XS 650 is also welcome with us.

Contact information:

Ringsted Motor V/ Per Jacobsen
Roskildevej 236
4100 Ringsted
Danmark
Tlf. +45 57617019
Mail: mc@ringsted-motor.dk
Www.ringsted-motor.com

Everything you need to keep your Yamaha classic fit and healthy!

You can get almost any spare part for your classic 70s or 80s Yamaha bike from us - either directly from our stock, or we will make every effort to track down the right match.

We have a comprehensive stock of:

New genuine Yamaha parts from old warehouse stock and current production

Reconditioned used parts and reproductions of parts that are no longer available

We have been dealing with Yamaha motorcycles for more than 10 years, and during this time we have not only accumulated a large stock of parts, but have also acquired the wealth of experience and knowledge – which we would be glad to share with you – that allows us to restore our classic machines and keep them on the road.

While it is true that we have a strong presence on the eBay sales platform, this doesn't cover our complete stock. Should you not be able to find a part you are looking for, simply give us a call or send us an email, and we'll see what we can do for you. We would be happy to make you an offer for multiple items, and because we are more flexible with our prices on our website, we can always compromise a little.

The tired kraut,   www.der-muede-bjoern.eu

The Yamaha XS 650 specialist
- *Parts* - *Engine rebuilding* - *Crankshaft rebuilding*
- *High performance parts*

Jerry van der Heiden
Chromiumweg 23
3812 NL Amersfoort
Netherlands

Tel: 0031 - (0)33 2862800
Fax: 0031 - (0)33 2861311

www.heidentuning.com

XS 650 Shop
Rüdiger Paustian

Importer / Distributor Reproduction Parts
Yamaha XS 650 Specialist
XS Performance Motorcycle Rims

Marga Faulstich Str. 6   24145 Kiel   Germany
Phone: +49 431 7197745   Fax : +49 431 7197746
Email : info@xs650shop

www.xs650shop.de   **www.xsperformance.de**

4 Tools

The XS 650 engine is a simple and logical built engine. So repairwork can also be done by people with little experience and with another profession than automotive technican. However, if two things - little experience and bad working conditions - come together, hardly a good result can be expected. The less experienced and skilled one is as a mechanic, the better the equipment should be. As described in some workshop manuals (not issued by the vehicle manufacturer) how to pull the rotor of the generator and how to loosen the sprocket nut without proper tools, certainly some rotors and engine housings have been destroyed.

By comparison with the possibilities of a professional workshop quite little equipment is needed. The necessary equipment, as described in the following, is enough to completely dismantle the engine, replace defective parts with spare parts and then reassemble the engine again. However, I consider it to be the absolutely necessary minimum. Someone who is working on something, with which he is not familiar, should be able to focus on the work alone and should not have to struggle with inadequate tools, poor lighting or an unheated workplace. While someone with a lot of experience and skills may be able to repair an engine "on the kitchen table", one who works on an engine for the first time should do this only under optimal conditions. If parts of such an old engine get damaged, it will get by far more expensive than it would have been if the work was done by an professional mechanic.

It is very important that the engine is thoroughly cleaned before disassembling it. This is best to be done with diesel fuel and a brush. However, you should have a sufficiently large tub to catch the running fluid. The cleaning of the engine can take two hours and more depending on the amount of dirt and the equipment. In particular, if the engine is not completely disassembled during repair, it is important that it is thoroughly clean externally to prevent dirt from entering the engine. If dirt from outside remains inside the engine, this will certainly lead to further damage.

For the disassembling and later assembly of the XS 650 engine a closed space with a workbench and usual tools such as a set of fork- and ring wrenches, a ratchet box, a torque wrench, a set of screwdrivers, middle sized pipe pliers, a string cutter, a locksmith hammer and a soft hammer are necessary. A compressor and an engine stand (Figure 4-1) should also be available. On the engine stand, the engine can be rotated by 360° around its longitudinal axis. Such universally applicable engine stands are offered by dealers on Ebay very cheap, so it is not worthwhile to build an engine stand by oneself, if the material has to be bought. To mount the engine on the engine stand, you only have to make an adapter.

Figure 4-1

If the engine housing is not to be opened and it should be only worked on the cylinder head or the clutch - so that it is not necessary to turn the engine to the side or to turn it completely - a simple self made engine stand as shown in figures 4-2 and 4-3 is sufficient.

Figure 4-2

The XS 650 engine is quite heavy and it can be handled only with difficulty by one person without suitable devices. The device shown in figure 4-4 can be used as an aid to lift the engine with the aid of a hoist and attach it to the engine stand. The device is bolted to the upper mounting points of the engine with the long M8 bolts. However, there is still a more suitable device for removing the engine from the frame, which is shown in the following chapter.

Figure 4-3

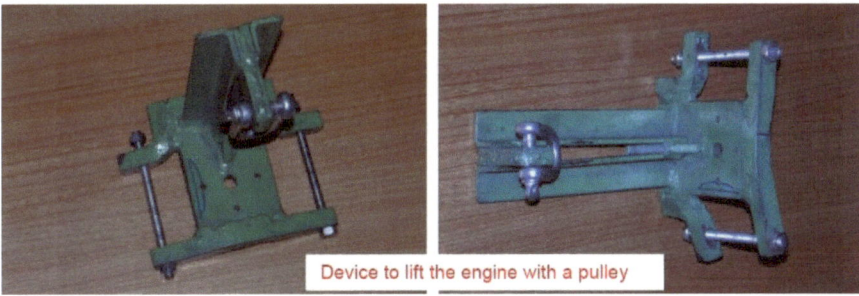

Device to lift the engine with a pulley

Figure 4-4

For the XS650 engine only one special tool is needed, that you cannot make by yourself without the means of a turning lathe. To remove the rotor of the generator, a special puller (figure 4-5) is necessary.

Puller fo the rotor of the generator

Figure 4-5

In principle, apart from the puller on figure 4-5, you do not need any special tools to disassemble the engine and then assemble it again, if you don't intend work that can only be carried out in engine repair workshops anyway.

However, the task can be made much easier, and there is less risk of damaging something if you build some of the tools described in the following. The tools are deliberately not described by means of technical drawings, but by means of photographs. You should improvise here and use material that is currently available. As a matter of experience, the time, which is required for the preparation of aids is saved later in the actual work.

A self-made tool as shown in figure 4-6, can be used to hold the sprocket to loosen the nut. The distance between the bolts was originally designed for a 17 sprocket. After grinding the bolts it also fits to an 18 sprocket.

Figure 4-6

Figure 4-7 shows a self-made tool for holding the clutch boss when the central nut is loosened. The relevant dimensions are shown in the figure.

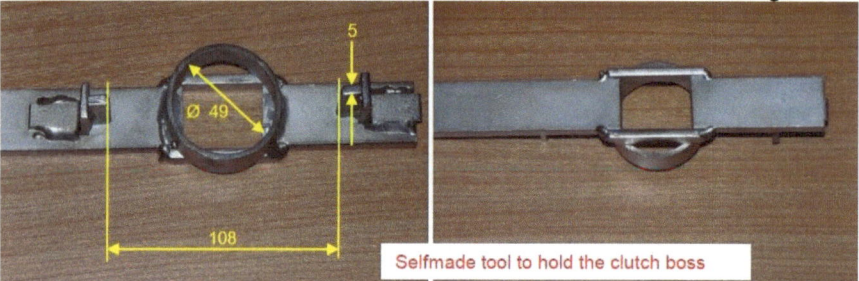

Figure 4-7

If the transmission shafts are also to be disassembled, one needs Seegerring pliers (figure 4-8). If you need such pliers only rarely, universal pliers with exchangeable tips, like the two shown on the left side of the picture are sufficient.

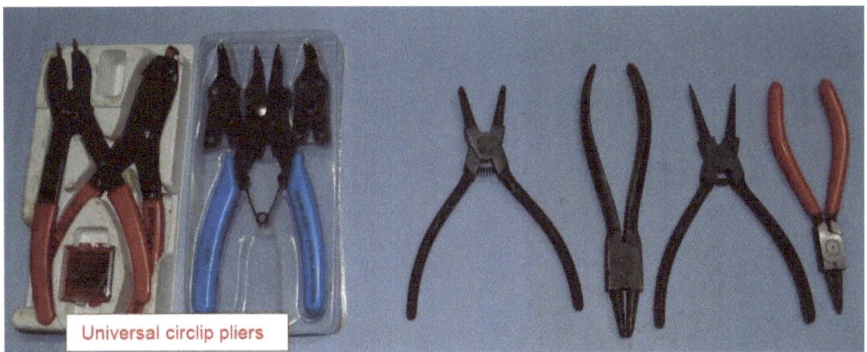

Figure 4-8

If the valves are to be removed, a valve spring press is required (Figure 4-9). If you buy a valve spring press, you should ensure that the inserts are large enough, since the valves of four-valve car engines, which are customary today, are smaller than the valves of the XS 650 engine.

Figure 4-9

The valves can be grinded with a device (figure 4-10). It can be powered by a cordless screwdriver and it changes the direction of rotation each time you slightly lower the pressure.

Figure 4-10

Piston ring pliers and a piston ring tensioner (figure 4-11 left) are not absolutely necessary. For mounting the pistons, the rings can also be tensioned with a commercially available hose clamp and a section of a can with a suitable diameter. For removal and assembly of the rings, plastic bands with a width of approx. 8 mm and a thickness of approx. 0.5 mm (figure 4-11 right) are sufficient.

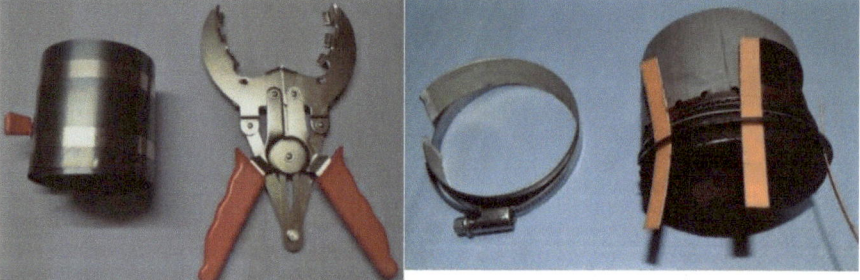
Figure 4-11

For the separation of the housing halves an approximately 20 mm thick round piece of wood and wooden wedges made from ash wood, as they can be bought for the assembly of doors in hardware stores, have proved their worth.

Before reassembling the engine, all parts must be thoroughly cleaned. Ideal is a parts washing machine (figure 4-12), as it can be bought quite cheap today. However, a complete engine does not fit into the parts washing machine shown in figure 4-12. After washing, the cleaning liquid should be allowed to drip for some time and then the parts should be dried with compressed air. Compressed air is also required to check whether the oil ducts are not clogged and have to be cleaned before assembly.

Figure 4-12

5 Removing the engine from the frame

The primary drive, the oil pump and the entire clutch including the shift shaft as well as the kickstarter mechanism can be dismantled when the engine is still installed in the frame. The engine must be removed from the frame in order to disassemble and repair all other components. The tank, both exhaust pipes, the horn and the footrests must be dismantled as preparatory work. Then the chain must be removed from the sprocket and pulled backwards as far as the swingarm.

It is not absolutely necessary to dismantle the carburetors, but as the space inside the frame is very tight, it is recommended to also remove the carburetors. The cables from the generator to the rectifier and the voltage regulator, from the contact breaker to the ignition coils and - if present - to the starter motor must be disconnected.

In figure 5-1, the mounting points of the engine in the frame are indicated by arrows. After all other bolts have been removed, finally the nut of the lower attachment (arrow mark) is released, but don't remove the bolt. Figures 5-2 and 5-3 show a device with which the engine can be lifted out of the frame without physical exertion by one person using a chain hoist. The suspension point of the chain hoist should not be perpendicularly above point where the chain is linked to the device, but rather outwards as (viewed from the frame) so that the engine can "swing" away from the frame. The distance between the suspension point of the chain hoist and the point where the chain is linked to the device should be as large as possible (high space). Thus the engine can be pulled out of the frame in a nearly horizontal movement.

Fixing points of the engine

Figure 5-1

First of all, the engine is lifted by means of a chain hoist (while pressing the handles of the device simultaneously down) so that the last bolt can be pulled out easily. If the engine is now lifted further and tipped slightly forward using the handgrips of the device, the engine will automatically swing out of the frame when the suspension point of the hoist has been selected outside the center of the frame.

First of all, the engine is lifted by means of the chain hoist (while pressing the handles of the device simultaneously down) so that the lower bolt can be pulled out easily.

Auxiliary device to remove the engine from the frame

Figure 5-2

Lower the engine on a suitable support

Figure 5-3

Figure 5-4 shows a simple engine stand on which the engine should be lowered. This can also serve as an engine stand if it is only planned to work on the cylinder head or the cylinders and the pistons. If the crankshaft is also to be removed or repair-work on the transmission is necessary, this engine stand is not suitable.

Figure 5-4

If the engine is now lifted further and tipped slightly forward using the handgrips of the device, the engine will automatically swing out of the frame when the suspension point of the hoist has been selected outside the center of the frame.

On the underside of the engine there are cap nuts, which are easily scratched when the engine is placed on a stone floor. The cap nuts serve the purpose to seal the studs of the lower engine housing half. If the "cap part" of the nuts is damaged oil can leak out. New cap nuts are not available as standard parts and are quite expensive as genuine spare parts. The dismantled engine should therefore never simply be placed on the ground but always on an engine stand, which can easily be made from angle irons, or it should be placed on a soft ground.

If the engine is only to be parked on the engine stand, two 400 mm long pipes with a diameter of approximately 40-50 mm are sufficient as shown in figures 5-3 and 5-4. Such an engine stand can easily be built. First drill holes into two short pieces of a flat iron bar. The short pieces then are welded to the pipes. A threaded rod is inserted through the holes in the flat iron bars and the lower mounting holes of the engine. Then, with the nuts tightened, a flat iron bar with a length of approximately 185 mm is welded between the pipes. This flat iron bar is then connected to the rear, lower engine mount. Since no one will buy material specifically for such an engine stand instead of using what is just available, I will not give a more detailed description. You should improvise here.

6 Disassambling the engine

In the following is described by means of photographs how to disassemble the engine. The diassembly begins with the cylinder head. If only repairwork on the transmission is planned, the work steps for the dismounting of the cylinder head, the cylinders and the pistons are omitted and one begins with the separation of the engine housing halves.

Basically, the disassembly of the engine can be started from both directions, from the bottom - the transmission side - or from the cylinder head. If only the clutch is to be repaired, it is not necessary to separate the housing halves. The camshaft can also be dismantled with the cylinder head cover removed without disconnecting the timing chain. The camshaft bearings can also be changed. To prevent that the engine housing halves warp, a certain sequence must be observed when loosening the screws that connect the housing halves.

The engine and all components must be absolutely clean during the later reassembly. Therefore, the engine should have been cleaned before disassembly. This prevents that dirt gets into the engine interior from the outside, especially from the area of the chain sprocket.

Engine on a selfmade assembly stand

Figure 6-1

Before starting to disassemble the engine, you should be sure that all the tools required for the planned repairwork are available, or you can make them yourself (chapter 4). If you disassemble an engine for the first time, you should plan at least one complete weekend as the time that will be required. This includes removing the engine from the frame, cleaning all individual parts of the engine and store them in a suitable order. You will need storage boxes, which can be bought in a hardware store, and many plastic bags (e.g. freezer bags), for the small parts. The more orderly the individual parts are deposited, the easier the subsequent assembly will be. If you are familiar with the function of the individual modules as described in chapter "2 Description of the function", the disassembly and, above all, the subsequent assembly will be much easier.

6.1 Disassembling the cylinder head

The sequence of dismantling - starting with the cylinder head - is described by means of photographs, which represent the individual steps. First, the governor assembly and the contact breakers are dismantled:

Figure 6-2

Figure 6-3

Figure 6-4

Figure 6-5

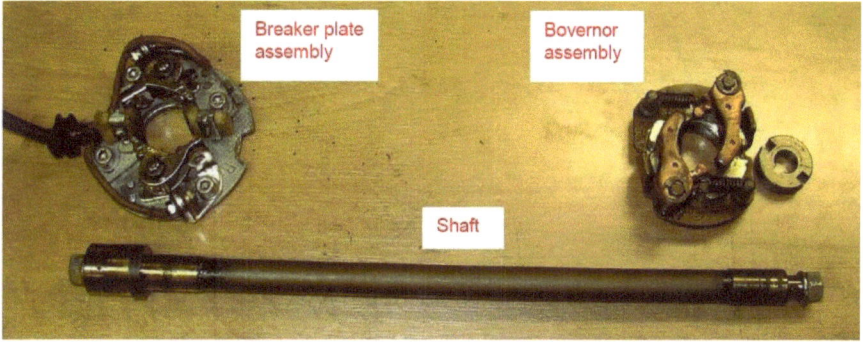

Figure 6-6

Then remove the oil delivery pipe and loosen the cylinder head screws. When removing the oil delivery pipe, first remove the upper banjo bolts so that the oil delivery pipe can not be twisted and damaged when the lower screw connection (Figure 6-8, right side) is released.

Figure 6-7

The sequence for loosening and tightening the cylinder head screws is shown in figure 6-9. Important: All screws must be loosened evenly, i. e. each screw is first loosened by ½ turns and then the other screws are loosened. In this case, the screw between the carburetor flanges, which can be seen in figure 6-11, must not be forgotten in order to avoid damage to the nut thread.

Remove the oil delivery pipe

Figure 6-8

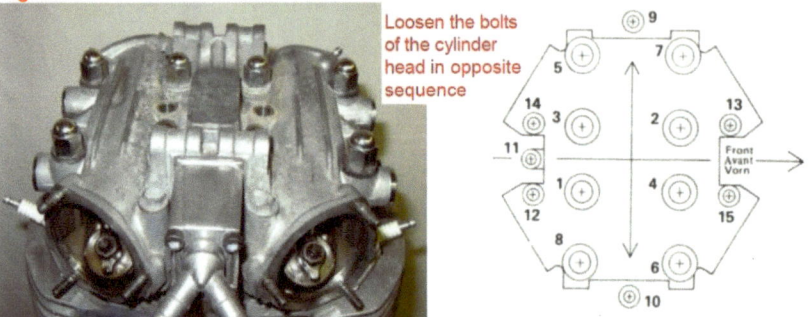
Loosen the bolts of the cylinder head in opposite sequence

Figure 6-9

Loosen the bolts of the cylinder head

The outer washers are rubberized

Figure 6-10

Loosen the bolts close to the spark plugs

Don't forget this bolt

Figure 6-11

Now the cylinder head cover can be removed and the timing chain can be separated by grinding off one bolt. If the engine is not to be completely dismantled, no chips must fall into the crankcase. If the engine is on an engine stand (figure 4-1), it is best to turn it 180° so that the timing chain is at the lowest point and no chips can fall into the crankcase.

Figure 6-12

Next, the lower part of the cylinder head is dismantled.

Figure 6-13

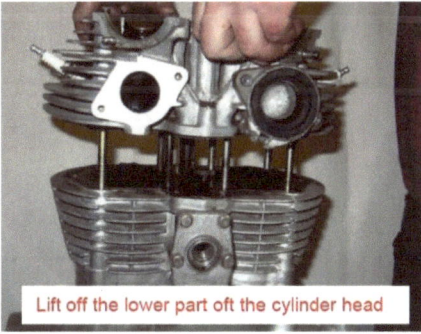

Figure 6-13

Then, the cylinder bank is removed and the pistons are freely accessible so that they can be heated. It is advisable to turn the engine to the side to avoid that the piston pin clips can fall into the crankcase. It is not necessary to dismantle the two piston pin clips of the individual pistons in the middle. If possible, a new piston pin clip should be used per piston.

71

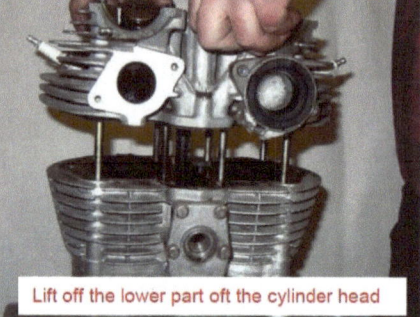

Figure 6-14

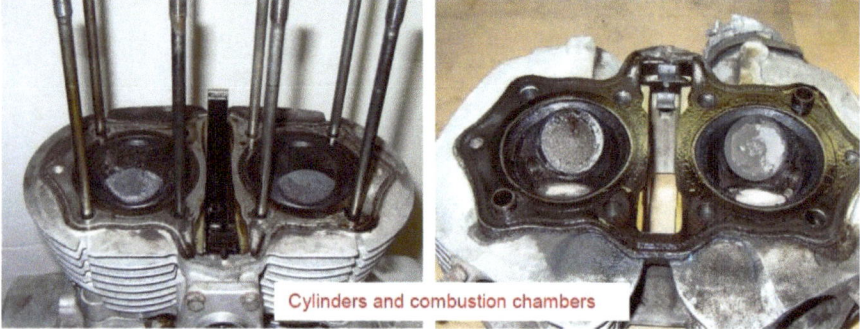

Figure 6-15

Figure 6-16

6.2 Dismantling the pistons

First, the piston pin clips must be removed. It is advisable to turn the engine to the side to prevent that they fall into the crankcase. Then, the pistons are heated so that the material expands and the piston pins can be pulled out as shown in figure 6-17.

Figure 6-17

6.3 Dismantling the generator

First remove the two long bolts of the stator and remove the stator. The centering pin must not get lost. The nut holding the rotor is released by hammering on a wrench as demonstrated in figure 6-19 (left). Then the rotor can be removed with a three-arm puller (figure 4-5) from its conical seat. **Never use a two-arm puller.** The woodruff key must not get lost.

Figure 6-18

Figure 6-19

73

Figure 6-20

6.4 Dismantling the clutch

First the right engine cover has to be removed. The bolts should be loosened crosswise. Caution, when removing the engine cover, make sure that the kickstarter mechanism remains in the engine housing. The kickstarter mechanism remains mounted. Pay attention to the thrust washer (figure 6-24).

Figure 6-21

The oil pump and the r. p. m. counter drive are located in the right engine cover.

Figure 6-22

Figure 6-23

Figure 6-24

Loosen the bolts evenly and crosswise - press the screwdriver against the spring tension – maintain the pressure during the first revolutions - otherwise the thread will break.

Figure 6-25

The individual parts of the clutch up to the clutch boss can then be dismantled. A tool (figure 4-7 and 6-28) should be used to hold the clutch boss when loosening the nut.

Figure 6-26

Figure 6-27

Figure 6-28

Figure 6-29
Removed clutch boss.

6.5 Disconnecting the engine housing halves

To separate the upper and lower engine housing halves, a total of 18 screws must be loosened in a certain sequence as shown in figure 6-36 on page 79. The screws must be loosened evenly so that the housing halves do not warp. One of the screws is located behind the clutch, so that the clutch must be removed before seperating the housing halves. In the vicinity of the housing screws there are indented numbers indicating the sequence in which they should be loosened or tightened. Depending on the condition of the housing, the numbers may be illegible. The housing screws are tightened again in the reverse order during the subsequent assembly.

Figure 6-30

Figure 6-31

To loosen:
Start with screw no. 18 loosen it ½ turns and proceed with the others. Then repeat the procedure. The separation of the housing halves requires a great deal of care and patience since the sealing surfaces, which are only sealed against each other by means of a very thin liquid seal, are very easily damaged. In the case of damaged sealing surfaces, reliable sealing can hardly be achieved.

Figure 6-32

Figure 6-33

Figure 6-34

Figure 6-35

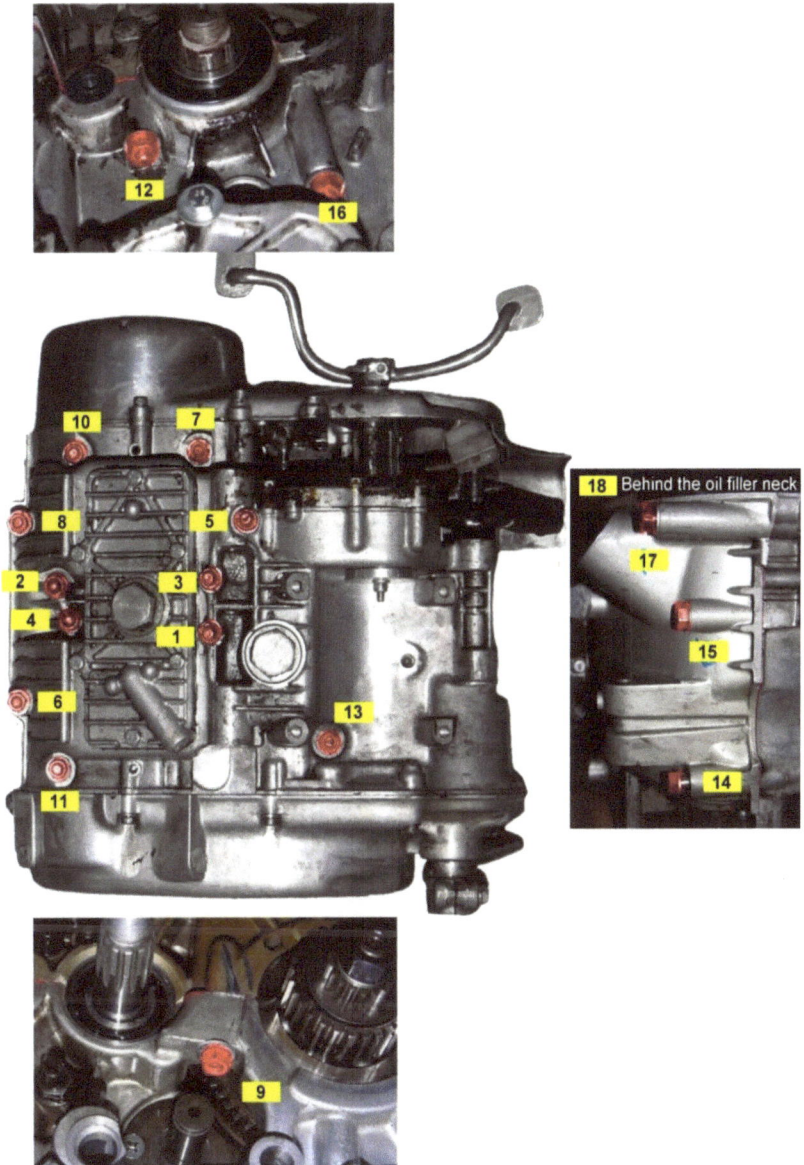

Figure 6-36

6.6 Removing the crankshaft and the transmission

The two shafts of the transmission can be easily removed from the upper housing half. If they are very tight, you can use a light plastic hammer. The crankshaft can be removed from the upper housing part in the same way.

Figure 6-37

Now the shift cam can be disassembled. First of all, the shifting forks must be disassembled. For this purpose, the cotter pins are removed first. They can not be re-used - so you just pinch them with pliers and pull the rest out.

Figure 6-38

The positioning of the cotter pins is important in the subsequent assembly, since a wrongly mounted cotter pin of the middle shift fork hinders the movement of the shift fork. More difficult is the disassembly of the pins which translate the rotational movement of the shift cam into a sliding movement of the shift forks, since the pins can not be grabbed with a tool like pliers. Push a new cotter pin into the hole on the end face of a pin. Then spread the cotter pin with a thin screwdriver blade as shown in the figure 6-39. Now you have a frictional connection between the cotter pin and the pin and can pull out the pin.

Figure 6-39

Now the spring (figure 6-40) must be released and the two screws for locking the retaining plate of the shift cam and the guide shaft of the shift forks can be removed. The shift cam and the shaft for guiding the shift forks can now be pulled out to the right side.

Figure 6-40

7 Assembling the engine

In the following it is described how to reassemble an engine, which was completely disassembled.

Figure 7-1

When describing the assembly of the individual modules, I will briefly review their function, but it is impossible within the scope of this book to list all the work steps or to graphically represent the complete assembly process. For the same reason I cannot explain general skills such as the assembly of shaft seal rings (Simmerrings), using paper seals, etc. Nevertheless, without understanding the funcion of the individual modules, it will be difficult to reassemble them properly.

So in order to avoide errors, you should have familiarized yourself with the function of the subassemblies as described in chapter "2 Description of the function". Also out of respect for the old technology, mistakes, which cause the destruction of spare parts, which are no longer produced, should be avoided. The assembly of the engine should therefore only be started by someone who is sure that he will master it.

Before you start to reasemble the engine, all parts should be carefully examined for their condition. The oil ducts in particular must be checked if they are clean. It must be sure that the fault, which has caused the reason for the repair was found and corrected, and not simply new parts are installed. If the reason for the defect is not found and elimited, the parts, that were replaced, will soon again be defective, if the actual cause of the failure - as for example a clogged oil channel - has not been cleaned. Before starting to reasamble the engine, check that all small parts, such as, for example, paper seals, shaft seal rings and standard parts such as screws, nuts and washers are available. However, it is by no means necessary to replace all seals after dismantling an engine.

In any case, however, the shaft seal rings on the left crankshaft stub and the shaft seal ring of the clutch push rod should be replaced as these have a retaining lug on the outer circumference and can therefore not be replaced without disassembling the engine housing halves. All other shaft seal rings can be replaced when the engine is installed in the frame, so that it is not absolutely necessary to replace them if they were not leaking before dismantling the engine. Paper seals can also be used again if they have not been torn. However, paper seals can also be produced very easily - and inexpensively - from sealing paper, as it can be purchased from the well-known accessory dealers. Sealing means (e.g. the red sealant from Dirko) are also required for the engine housing halves, for which a paper seal is not provided.

On paper gaskets for covers, which have to be removed for maintenance purposes, a sealing agent like the blue Hylomar should be applied. By this way, the paper gaskets do not stick to the metal surface and do not tear when the covers are removed, so that the gaskets can be reused. Often, the mistake is made that the processing instructions of Hylomar are not observed and that it is applied too thickly. This must be avoided in any case because overflowing sealing compound can clog the oil channels.

In order to secure the piston pins, new clips are needed, when the pistons have been dismantled. New cotter pins are needed when the shift forks have to be reassembled.

I strongly recommend to use an engine stand as described in chapter "4 Tools", especially if the disassembly without has already caused problems. It depends, of course, on the skill of the individual, whether he may also do without an engine stand. However, anyone who works on an engine for the first time should do it only under optimal conditions.

New parts will not always be available when an engine is reassembled. The components of the actual engine, such as pistons, bearings, the crankshaft and the camshaft are replaced after a defect by new parts or parts refurbished by a specialist company.

After a defect of the transmission it is almost always necessary to fall back on used spare parts which must be carefully checked before installation, since new parts are usually not available. To what you have to pay particular attention, I will point out in the individual chapters. Examples can also be found in chapter "8 Typical damage".

7.1 Shifting mechanism, gear box and kickstarter

When reassembling the engine, start with the upper housing half, into which the crankshaft is inserted.

Figure 7-2

Care must be taken to ensure that the centering pins, which fix the crankshaft bearings in the engine housing, are correctly positioned. The centering pins position the outer rings of the crankshaft bearings in such a way that the engine oil can enter the bearings. You can not do anything wrong here, if the bearings are not correctly positioned, the housing halves cannot be closed. When the crankshaft is positioned in the upper engine housing part, the timing chain can be placed loosely on the toothed wheel of the crankshaft and be secured by means of wire from the opposite side.

Figure 7-3

The gearshift forks are put into the upper engine housing half in the correct order and orientation. Check that the guide bar is absolutely clean, straight and free from scratches and that the plug on the side which faces the clutch isn't missing. Then push the guide bar through the hole in the right side of the upper engine housing part, through the small holes of the shift forks and finally into the blind hole on the left side of the engine housing part (figure 7-3).

Oil flows through the holes in the guide bar for the lubrication and cooling of the gear wheels and to the bearings on the clutch side of the transmission shafts. Contamination of the guide bar can hinder the flow of oil, and thus cause further damage.

Next, the shift cam is pushed through the large bore in the right side of the upper engine housing half, and the three guide pins of the shifting forks are inserted into the bores of each shift fork. The guide pins are secured by means of cotter pins (3.2 x 32 DIN 94), as shown in Figs. 6-38 and 6-39 on page 80. Make sure that the cotter pins do not interfere with the movement of the shift forks.

Figure 7-4

Now the shift cam can be fixed by means of the half-moon-shaped plate (stopper plate) and two hexagonal bolts, which are again secured with a securing plate or a liquid screw retention (figures. 7-5 and 7-6).

Now the star shaped gear lock wheel (red arrow mark on Fig. 7-5) and the pin ass'y for neutral (figure 7-6) is mounted. If possible, a new spring should be used if you are not sure whether the existing spring has lost tension. The same applies to the spring for the gear lock wheel. Even if it seems that there is still enough tension, it is better to use a new spring. If there were previously problems to shift the transmission in neutral while the engine was warm, then the reason was also the interaction of these two springs.

The spring in figure 7-6 to the right keeps the shift cam in the idle position while the spring on figure 7-6 left holds the shift cam in the positions corresponding to the shift state of the individual gears.

If the pretensioning of the spring of the star shaped gear lock wheel is too small, the gear pairs of the individual gears are no longer held securely in their positions on the respective shaft of the transmission. There is the risk that the corresponding gears unintentionally disengage. If this happens more frequently, it is no longer simply repaired with the replacement of the springs because the shifting claws and the corresponding recesses of the opposite gear wheel get damaged each time when shifting.

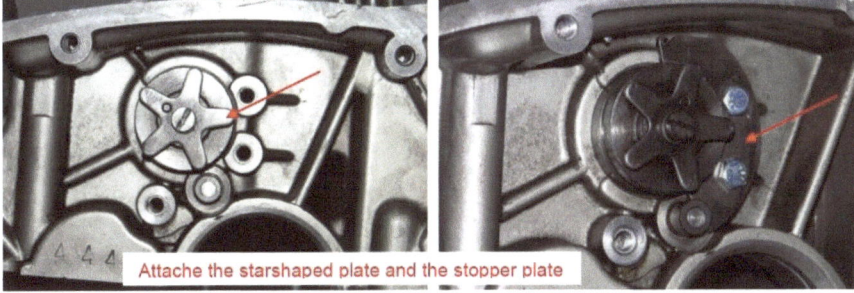
Attache the starshaped plate and the stopper plate

Figure 7-5

Attache stopper lever ass'y and secure it with the tension spring

Pin ass'y for neutral

Figure 7-6

Functional check:
Now, you should try to turn the star shaped gear lock wheel to check if the shift cam can be turned freely and whether the shift forks can be moved freely.

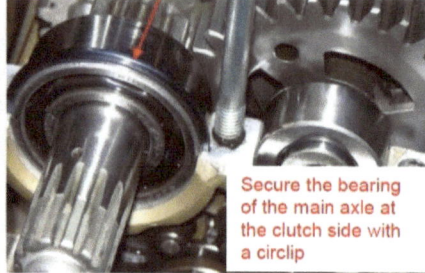

Figure 7-7

Mounting the transmission is easy because the gear shafts are simply inserted (figure 7-7). In this case, ensure the correct positioning of the retaining rings on the clutch side of the transmission input shaft (arrow marking in figures 7-7 and 7-8).

Figure 7-8

Figure 7-9

Figure 7-10

Since hardly a new transmission is installed, it should be checked before the installation whether all gear wheels can be turned freely and can be moved freely lateraly on the respective shaft. **Above all, it must be checked whether the shift claws and recesses in the end faces of the counter-gears are not worn out.** An example of how worn-out shift claws look, can be found in chapter "8 Typical damage".

The retaining ring of the fixed bearing (circlip) on the pinion side of the transmission output shaft consists only of one half and must be inserted so that it engages with both housing halves (figure 7-8).

A collar (red arrow mark on figure 7-9) is then pushed onto the toothing of the transmission output shaft. The end faces of the collar and its outer diameter must be absolutely clean and free from scratches in order to ensure a secure seal. If the collar has already developped a slight groove, as shown in the example in figure 7-9, the collar can be installed in such a way that the sealing lip of the shaft seal does not run on the groove. Next, the shaft seal ring (simmerring) is pushed onto the output side of the transmission output shaft, and the shaft seal ring of the clutch push rod is positioned in the opening provided for this. Since the kickstarter mechanism (figure 7-10) has usually not been disassembled, it only needs to be inserted into the bore provided in the lower engine housing part. The spring is pretensioned by at least one turn before insertion.

<u>Functional check:</u>
Now is the last opportunity to check the function of the shift mechanism and the transmission before closing the engine housing halves. By turning the star shaped gear lock wheel, the shift cam must be allowed to rotate with some effort, but still by hand, with the shifting forks moving in the axial direction on the shift cam and the guide bar. The shift claws of the individual gear wheels must be fully engaged in the recesses of the counter-gear wheels.

7.2 *Closing the engine housing halves*

When it is certain that the transmission and the shifting mechanism are working properly, the engine housing can be closed. Before assembling the engine housing parts, the shaft seal ring (Simmerring) must be slipped onto the crankshaft stub on the generator side of the crankshaft (figure 7-14).

Since the bearing seats of the crankshaft and the transmission shafts are in each case half in the upper and the lower housing parts, no paper gasket can be used here for sealing purposes. The sealing compound (e.g. Dirko red) must never be applied thicker than shown in figure 7-11. To apply the sealing compound, first clean the sealing surfaces with brake cleaner and then apply the sealing compound with a fingertip while simultaneously spreading it evenly and, above all, thinly on the sealing surface. Excessive sealing compound on the inside must be completely removed before joining the engine housing halves (e.g. with a cotton swab).

Lower crankcase with sealant | Put the lower crankcase on the upper one

Figure 7-11

Gap between the crankcase halfs

Figure 7-12

Clutch side of the crankcase

Crankshaft bearing with circlip on the clutch side

Figure 7-13

Gear shafts with oil seals on the left hand side

Crankshaft bearing with oil seal on the left side

Figure 7-14

Gap at the front of the engine

Before applying sealing agent to the sealing surfaces, check whether the engine housing halves can be closed easily. If not, check the correct fit of all parts, especially the crankshaft bearings.

Caution: Excessive sealing compound can clog oil channels!

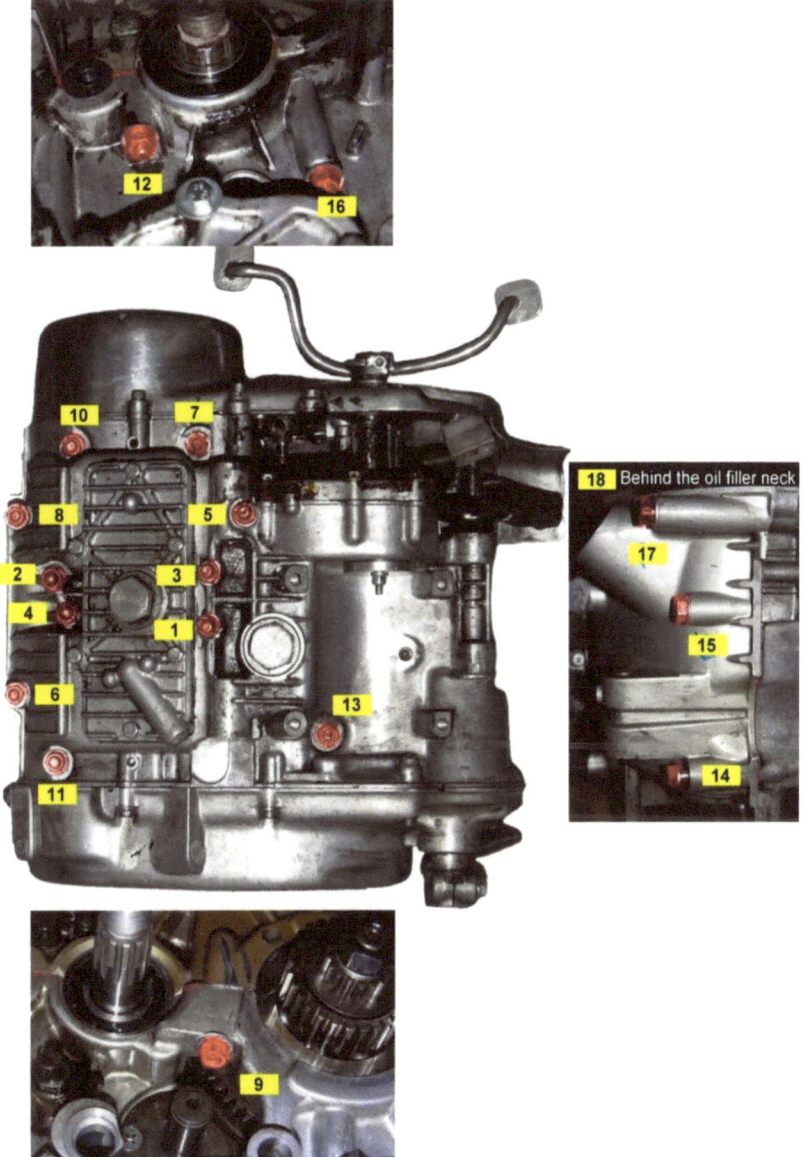

Figure 7-15

The figures 7-16 and 7-17 show the shaft ends of the crankshaft as well as those of both transmission shafts and the sealing ring of the clutch push rod. Because of a retaining collar the sealing rings of the clutch pressure rod and that of the crankshaft (figures 7-13 and 7-14) can only be replaced when the engine is removed from the frame and the engine housing halfs are separated. Before assembling the engine, these two shaft seal rings (Simmerrings) should therefore always be replaced. All other shaft seal rings can be replaced when the engine is built-in and can therefore be reused. The screws of the engine housing must be tightened uniformly and in a prescribed order (spirally from the inside outwards).

7.3 *Installation of the shift shaft and the clutch*

7.3.1 The shift shaft

The shift shaft is inserted from the clutch side of the engine (right) into the provided bore in the lower engine housing part.

Insert the shift shaft from the left hand side

Secure the shift shaft with a washer and an e-clip

Figure 7-16

The toothing of the shift shaft, with which it is linked to the shift lever, can easily damage the sealing lip of the shaft seal ring. This can be prevented by attaching adhesive strips (e.g., tesafilm) to the toothing. Nevertheless if the shaft seal has been damaged, this is no big problem. The shaft seal ring can be easily replaced when the engine is installed. On the left hand side of the engine the shift shaft is then secured in the axial direction with a disc and an E-clip. Finally, the upper fork of the arm welded to the shift shaft must be connected to the star-shaped disk of the shift cam.

Connect the shift shaft with the shift cam

Figure 7-17

The pins of the star-shaped disk must be located between the hooks as shown in figure 7-17. If they are not in the middle, this can be corrected by the eccentric screw.

7.3.2 The clutch

Next, the clutch is mounted as shown in the figures 7-18 - 7-29. Figure 7-18 shows the sequence of the washers and bearings from the inside to the outside by means of characteristic colors.

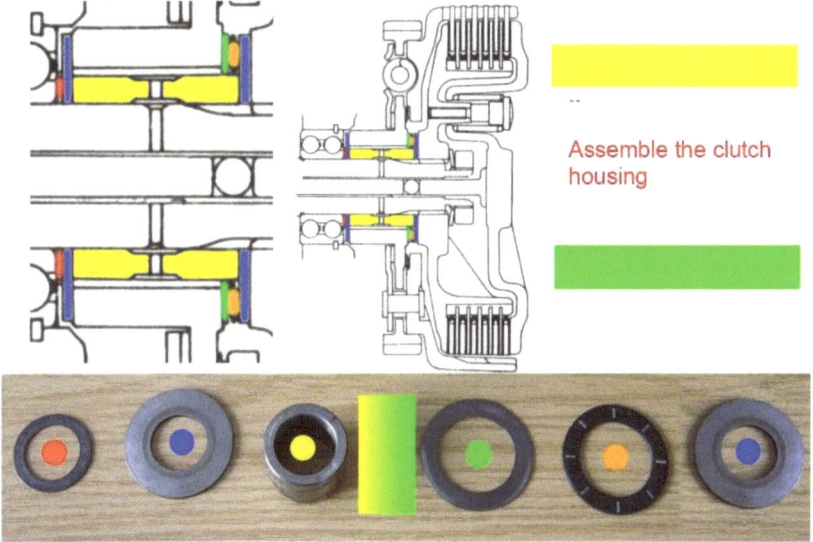

Figure 7-18

The yellow/green colour gradient symbolizes the clutch housing. In the following pictures, the assembly of the individual parts of the clutch is documented in the order of the work steps. The clutch housing and the clutch boss have markings, which must be aligned with each other.

Figure 7-19

Figure 7-20

Figure 7-21

Figure 7-22

Figure 7-23

Insert the clutch boss
Figure 7-24

Assemble washer, belleville spring and nut

Tighten the nut (75 – 80 Nm) using selfmade tool to hold the clutch boss
Figure 7-25

Insert the clutch discs - rounded side to the outside

Figure 7-26

There are no special features that need to be taken into account when installing the clutch, except for the alignment of the clutch plates (figure 7-26) and how to attach the compression springs (figure 7-27/8). To hold clutch housing when the central nut is tightened, use a suitable tool such as it was used for dismantling (figure 7-25). Lubricate everything well during assembly.

Place the pressure plate and install the clutch springs

Figure 7-27

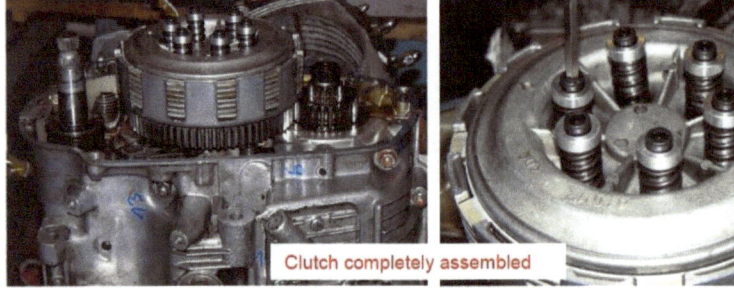

Figure 7-28

Figure 7-29

Tighten screws of the clutch springs evenly and crosswise - press the screwdriver against the spring tension - keep pressure during the first rotation of the screwdriver - otherwise, the threads will break.

7.4 Install the right side crankcase cover

Before installing the right engine cover make sure that the thrust washer on the kickstarter shaft (figure 7-29) is not missing. The oil pump, which sucks the engine oil from the oil sump in the lower engine housing part, is located in the right engine side cover, with an oil filter arranged in the direction of the oil flow behind the oil pump.

The oil pump is driven by means of a pair of spur gears by the crankshaft, while the tachometer drive is driven by means of a worm drive. Because of the engagement of the spur gears and, of the oil passage guided through the sealing surface, an exact positioning of the right crankcase cover is necessary, which is ensured by means of dowel pins. Since the engine oil - unlike with the sealing surface between the upper and lower engine housing parts - is constantly in touch with the sealing surface of the right crankcase cover even when the engine is stationary, a paper seal with a thickness of 0.5 mm is provided.

The friction plates of the clutch, which are concealed by the right crankcase cover, are wear parts, which can be replaced without expensive disassembly. Therefore it should be prevented, that the gasket gets torn when the right right crankcase cover is removed. As mentioned before, an agent like the blue Hylomar is a suitable means to prevent that the gaskets sticks to the sealing surface and gets torn when the crankcase cover is removed. When changing the friction plates of the clutch, the engine oil can remain in the engine when the motorcycle is tilted to a left-leaning position. If the engine oil remains in the engine, any residue from a possibly torn seal, which must be scraped off the sealing surfaces, should not get into the oil.

Figure 7-30

The sealant must never be applied thicker than shown in Figure 7-30, especially close to the oil passages.

Tearing of the gasket can be prevented with a "seal enhancer", e.g. the blue Hylomar, but - if this is not applied correctly - there is a danger that the oil channels will be clogged. Before applying Hylomar to the sealing surfaces, you should first check whether the engine cover can be closed without exerting a force. If this is not possible, you must check if all parts fit properly.

Before applying Hylomar the sealing surfaces, they must first thoroughly be cleaned e.g. with brake cleaner. Then can be applied to the sealing surface and spread with a fingertip thinly and evenly. On the inside of the engine housing, seal enhancer must be thoroughly removed before joining the crankcase cover. Once you are sure that the crankcase cover fits without tension, all screws are tightened evenly and crosswise. In the original workshop manual, a torque of 20 Nm is specified, which is easily exceeded by an Allen key when tightened "by feel". The threads in the upper and lower engine housing parts are often damaged if the crankcase cover has been removed already several times.

Repairing a damaged thread is particularly difficult with the thread in the lower engine housing part below the kickstarter shaft. In the case of undamaged sealing surfaces, a pressure, which can be achieved with cross-slotted screws, is sufficient for sealing. Not only who looks for originality should therefore perhaps reverse an "improvement" - the conversion to allen head bolts. In the case of damaged sealing surfaces, however, a higher pressure for sealing does'nt help. A suitable means is to machine the surface or use a suitable sealant, which fills smaller unevenness.

Place the cover together with the gasket and tighten the screws evenly crosswise

Figure 7-31

Tighten the screws evenly and crosswise - Attention: Do not forget the dowel pins!

7.5 Clutch-pushrod and the drive sprocket

The clutch push rod transmits a thrust movement from a fixed component - the push lever assembly - to the clutch plate as a rotating component. There are one-piece and two-piece clutch push rods, the second being the original spare part. If the two-piece clutch push rod is used, a ball (ø 8 mm) is first pushed through the sealing ring of the clutch push rod (arrow mark on figure 7-32). The longer part of the two-piece clutch push rod is then inserted, and then again a ball. Finally, the shorter part of the clutch push rod is inserted.

Figure 7-32

The installation of the drive sprocket requires some care. The sprocket must be pushed as far along the teeth of the transmission shaft so that the toothing is still slightly protruding, as it can be seen in figure 7-33.

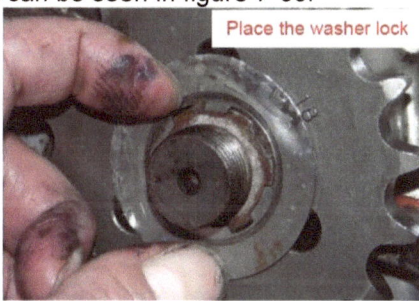

Figure 7-33

Figure 7-34 shows how the nut is mounted so that its collar engages over the toothing of the transmission shaft.

Figure 7-34

To tighten the nut (100 - 120 Nm), use a tool as described in the chapter "4 Tools" (figure 4-6). **In no way, as described in some manuals, you should block the sprocket by clamping it with the clutch push rod by using the drive chain.** Finally, the securing plate is bent over.

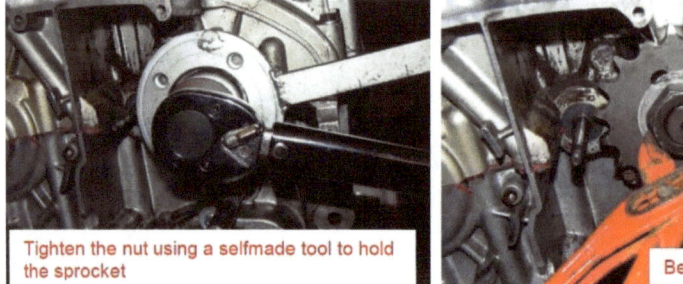

Figure 7-35

7.6 *The generator*

The generator consists of two parts: The rotor, which is mounted on the conical left crankshaft stump and is secured with a woodruff key and with a nut with a fine thread. The stator, which is bolted to the engine housing.

First, the woodruff key is inserted into the its seat on crankshaft stub.

Figure 7-36

Then the rotor is placed, followed by a lock washer and a nut.

Figure 7-37

Before the nut is mounted, apply liquid bolt retaining compound to the thread of the crankshaft. Since it is difficult to hold the rotor when the nut is tightened, the best way to tighten the nut is to use an impact wrench, as shown in figure 7-38.

Figure 7-38

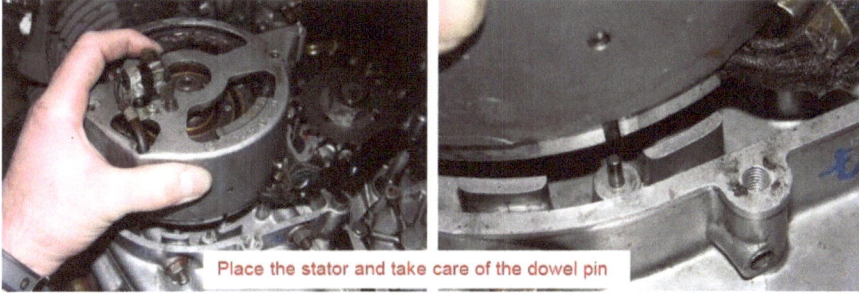

Figure 7-39

Figure 7-40

Figure 7-41

When the rotor is mounted, the stator follows. Take care that a pin in the engine housing engages into a gap provided for it in the stator housing. Then the cable is laid as shown in the figures 7-40 and 7-41, and the chain guide of the drive chain being mounted last.

7.7 *The pistons and cylinders*

The assembly of the pistons and cylinders is much easier if you turn the engine to the side, as shown in the following pictures.

Turn the engine to the side

Place the gasket and secure the timing chain

Figure 7-42

Insert the piston into the bushings, and pay attention to the arrow mark (front) on the piston. The front chain guide is already assembled and has to be adjusted to the timing chain.

Place the O-rings and insert the piston pins from outside

Figure 7-43

No special tools such as piston ring pliers or piston ring tensioners are necessary to mount the pistons and cylinders (Fig. 4-11).

With some skill, the piston rings can also be mounted and the pistons be inserted into the cylinders by simpler means. The gap of the piston rings should be offset by approximately 120°. The pistons are pushed into the cylinders so far that the bore for the piston pin remains free. Before the cylinders with the pistons already partly inserted are placed on the engine housing, the gasket is also placed (figure 7-42) and the seal rings of the cylinder liners are installed. The sealing rings are pressed into the groove before the cylinder block is fitted. Then the engine is turned to the side and the cylinders are pushed onto the stud bolts.

| Assemble the rear chain guide | Push the cylinder onto the stud bolts |

Figure 7-44

The timing chain is then secured with a piece of wire and pulled through the shaft between the cylinders. The connecting rod's bores are aligned with the bore for the piston pins and the piston pins are pushed through. For this purpose, it may be necessary to heat the pistons so that they expand and the piston pins can be pushed in more easily. When the piston pins are completely inserted, the retaining rings for the piston pins on the outer sides of the pistons are mounted.

Push the connecting rod eye into the piston and insert the piston pins into the connecting rod eyes

Figure 7-45

If possible, new retaining rings should be used. It is also advisable that the engine is turned to the side when the retaining rings are fitted to prevent that the retaining rings are dropped into the crankcase.

After the cylinders are lowered down to the gasket, the front guide bar has to be aligned. The long end of the guide bar belongs downwards. Then the pistons are brought into the TDC position. The ends of the timing chain are laid forward and rearward and are - e.g. with a piece of wire - secured. Now the cylinder head can be mounted.

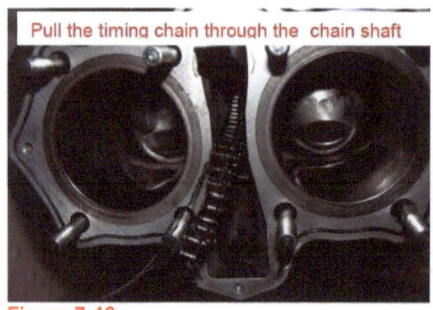 

Figure 7-46

7.8 Cylinderhead and camshaft

Figure 7-47

To mount the cylinder head and the camshaft, the engine is set upright again. If the cylinder head has been disassembled, a standard valve spring press (figure 4-9) is needed for the assembly of the valves. A valve spring press can be bought at a spare part- and accessories shop for approximately 50 € or it can be borrowed from a friendly car workshop.

The inner and outer valve springs are installed in a certain orientation.

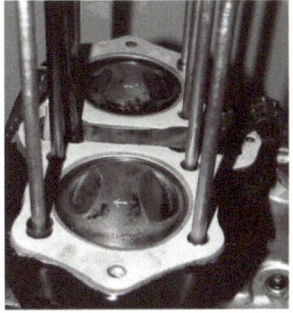

Place the cylinder head gasket

Figure 7-48

Assemble the cylinder head

Figure 7-49

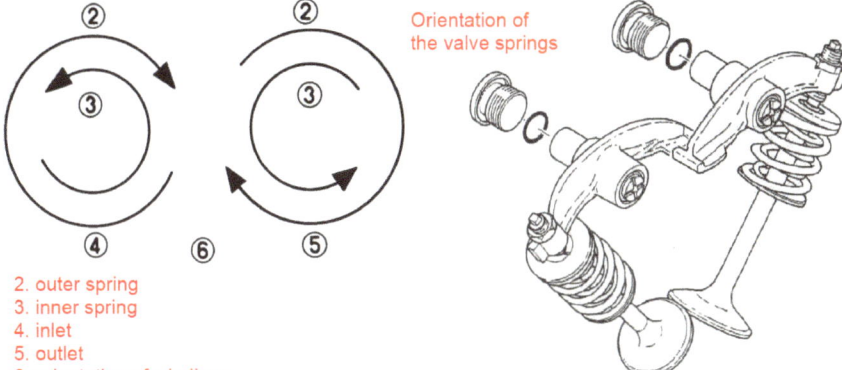

2. outer spring
3. inner spring
4. inlet
5. outlet
6. orientation of windings

Figure 7-50

Place the valve stem oil seals — Insert the valves

Figure 7-51

If necessary, new valve stem seals are fitted on the valve guide bushings. This is always recommended after a longer time in service, after which a removal of oil carbon or even a grinding of the valves was necessary. The valve stem seals can only be replaced when the engine is removed from the frame.

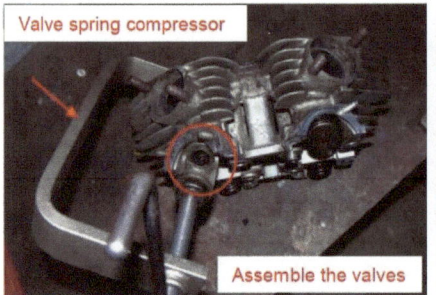

Figure 7-52

Cylinder head with fully assembled valves
Figure 7-53

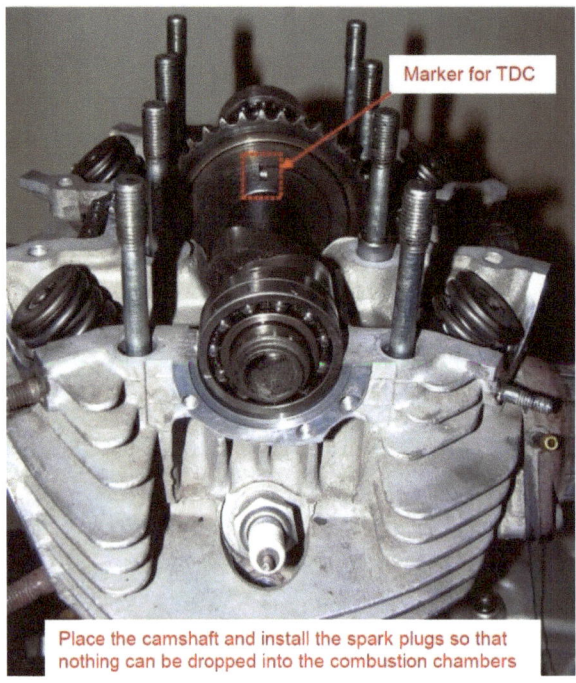

Figure 7-54

7.8.1 Riveting the timing chain

In principle, the already riveted timing chain can also be placed on its sprocket on the crankshaft before the crankshaft is inserted and can be assembled together with the crankshaft. The camshaft must then be installed before of the chain tensioner is assembled. The camshaft can be pushed under the raised timing chain. If the timing chain is not yet riveted, proceed as shown in the following figures:

Preload the cylinder block with two nuts, otherwise a new timing chain will be too short to be closed

Figure 7-55

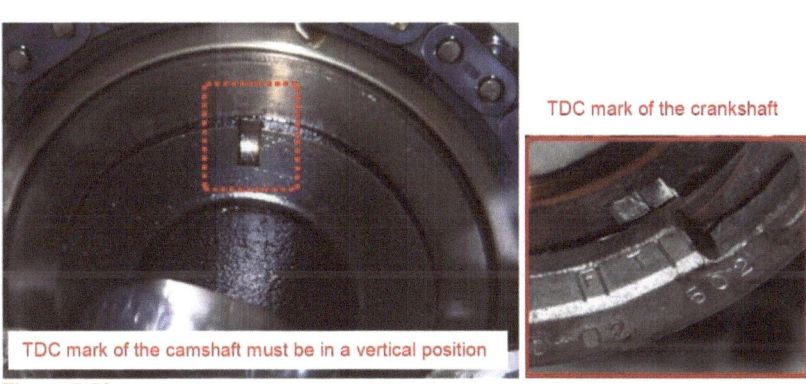

TDC mark of the crankshaft

TDC mark of the camshaft must be in a vertical position

Figure 7-56

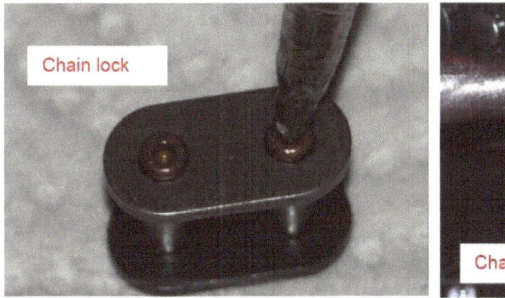

Chain lock

Chain lock - left rivet already riveted

Figure 7-57

105

Use a heavy hammer as a support

Figure 7-58

7.8.2 The cylinder head cover

The four camshaft bearings must be located as far inwards as possible. This is done by pulling the bearings together by means of a threaded rod and two hexagonal nuts while the cylinder head cover is being mounted and the cylinder head bolts tightened. To do this, use an M 12 threaded rod, as you can buy it in a hardware store. The nuts, which are used to pull the bearings together must not be too much tightened, otherwise the camshaft bearings get damaged.

The bearings should be only **fixed** in their innermost position, and not be pressed against each other. Therefore, one should use a threaded rod M 12 and no thinner one because it would be too elastic.

Hold the camshaft bearing in the inner position

Figure 7-59

The cylinder head bolts must be tightened in the same way as for dismantling in a sequence as shown in figure 7-60. The M 10 x 1.5 nuts are tightened first and then the M 8 bolts.

 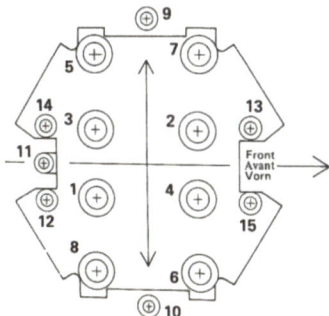

Install the cylinder head cover and the oil delivery pipe   The cylinder head bolts must be tightened in a sequence as shown

Figure 7-60

Before mounting the covers for the governor and the breaker plate, check whether the camshaft bearings are positioned sufficiently deep in the housing of the cylinder head. To do this, measure the distance between the areas indicated by the red and green arrows on figure 7-61 using a caliper with a depth gauge. The thickness of a paper gasket of about 0.5 mm is added to the measured value.

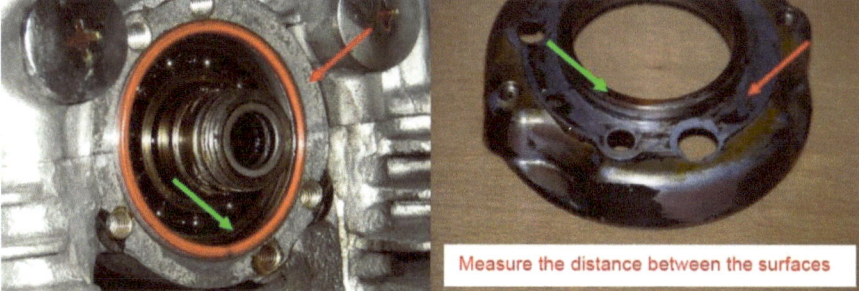

Figure 7-61

The distance to the cylinder head must be larger than the distance on the covers. If the distance on the cover is larger than on the cylinder head, the paper seal is not pressed and a gap remains, which prevents a proper sealing. If the bolts are tightened too much, damage can occur as shown in figure 7-61 above. The region of the cover, which is close to the bolt head, then breaks out. Reassemble governor and the breaker plate in the reverse order as shown in figures 6-2 to 6-5 in chapter "6 Dissassembling the engine".

7.9 *The timing chain tensioner*

The timing chain tensioner mechanism has been changed several times during series production so that, as far as the counter nut of the tensioner bolt is concerned, there are different variants of the tensioning mechanism, although the mode of operation remained the same.

Assembling the timing chain tensioner

Figure 7-62

It is also possible that e.g. a provided damper disc has not been installed, or the spring force has declined with time. The description in the original workshop manual, according to which the pin is to be flush with the end face of the hexagon within the tensioning mechanism, as shown in figure 7-63, can therefore only be a reference value.

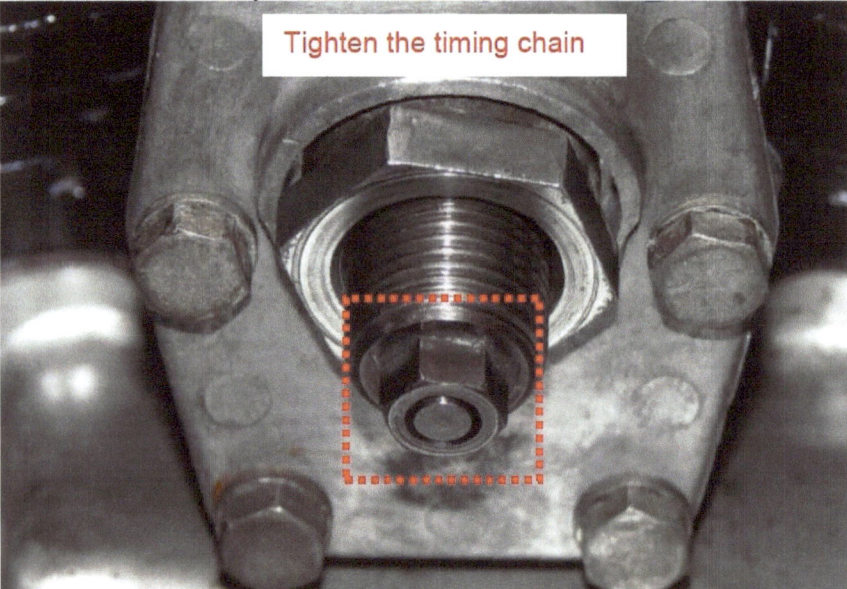

Figure 7-63

It is better to tension the timing chain by turning the pin while simultaneously touching the end of the pin with a fingertip. If the pin still pulsates slightly, the timing chain is properly tensioned. In particular, a too much tensioned timing chain can cause damage.

In case of doubt, you should rely more on your feeling than to tighten the timing chain too much. To adjust the tension by turning the hexagon you should use a short wrench or turn the hexagon only by hand. With a long ring-wrench you have too little feeling and there is the risk of tensioning the timing chain too much.

7.10 Concluding work

If the engine has been reassembled as far as previously described and the valves and the ignition are set, the engine can be installed in the frame.

Before you start the engine for the first time, you should be convinced that the oil circuit is working, but a complete test is not possible with the engine installed. The fact that the oil pump works at all can be checked by removing a valve cover on both sides and actuating the kickstarter when the spark plugs are turned out. After a while oil must arrive in the cylinder head - but you need some patience and have to kick long enough.

How to handle a new engine or an engine that has been overhauled, there are many opinions. Modern engines do not need to be conditioned anymore. This is because today's production methods are able to produce better surface finishes with small tolerances, which no longer necessitate that surfaces "get used" to each other.

In the case of the XS 650 engine, after a basic overhaul of the engine, only newly machined surfaces are formed between the cylinder linings and the pistons. However, if new oversize pistons have been installed, it is no exaggerated caution if you do not require the engine to perform at full speed in the first time for example over a distance of 400 miles.

If only components of the transmission or bearings have been replaced, a special conditioning is not necessary.

8 Typical damage

Damage to the engine and transmission occurs due to normal wear and tear in operation, due to lack of maintenance as well as faulty maintenance or repair work.

All rolling bearings, such as the bearings of the crankshaft, those of the camshaft and those of the transmission shafts, as well as the tooth flanks of the gear wheels and the pistons and cylinder linings are subject to a certain wear which can not be prevented even by regular maintenance. Except for the worn-out gear wheels, which have to be replaced by original gear wheels and which are not unlimited available, the XS 650 engine can be reworked several times.

8.1 Engine

8.1.1 Crankshaft

The main- and connecting rod bearings of the crankshaft are subject to constant wear during normal operation. If the wear exceeds the limits stated in the original workshop manual on page 27, the bearings should be replaced. The original workshop manual also describes how the wear is to be measured.

The crankshaft is an assembly of individual parts, which are joined by pressing together. These individual parts, such as the crank pins and the counter weights, can get twisted relative to one another.

No matter what kind of damage occurred, the crankshaft can only be repaired by the means which are usually only available in engine reconditioning workshops. In the chapters describing the disassembly and assembly of the engine, the description was therefore limited to the removal and installation of the crankshaft.

8.1.2 Pistons

The description of a damage such as it can occur on the pistons is also omitted here since there is no damage occurring exclusively on the XS 650 engine, which is typical for this engine.

Worn or destroyed pistons should not simply be replaced with oversize pistons. One should first look for the cause of the damage, for which there may be several causes besides normal wear. If the causes are not elimited, the same damage will also occur in the short-term again.

Information with illustrations of typical piston damage and a description of its causes can be found on the websites of piston manufacturers and in the brochures, which can be downloaded from those websites.

8.1.3 Valve drive, timing chain, tensioning rails

In figures 8-1 to 8-3 on the next page, the front and rear chain guides are shown in a worn condition.

Figure 8-1

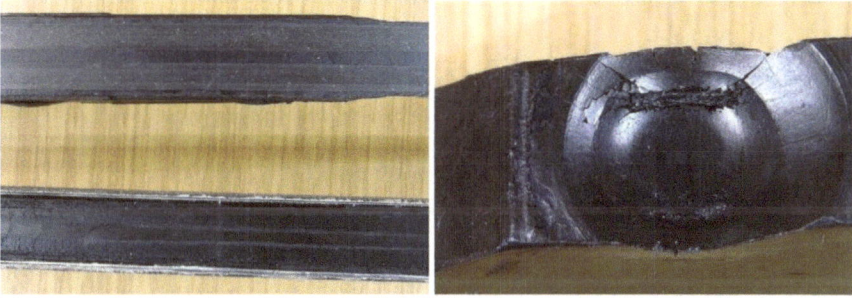

Figure 8-2

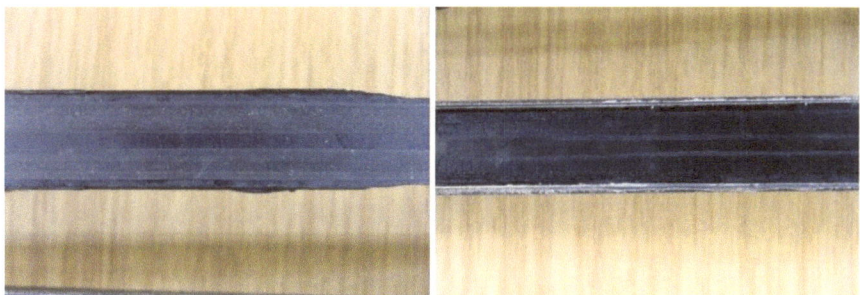

Figure 8-3

The figures 8-2 and 8-3 show the wear pattern in detail. Figure 8-2 also shows the spherical dome on the back of the rear chain guide, into which the timing chain tensioner engages.

Excessive wear on chain guides and on the timing chain itself occurs both when the timing chain is tensioned too loose or too tight. A too loose tensioned timing chain tends to rock, while too much tensioning of the timing chain exerts too much force on the sliding surfaces of the chain guides and the spherical dome, which leads to similar damages.

8.1.4 Camshaft

Damage to the camshaft occurs due to oil deficiency or by a too large valve clearance. A too large valve clearance is noticed due to excessive noise development and generally it is adjusted correctly before damage occurs. Figures 8-4 and 8-5 show a camshaft, the cams of which have been under excessive heat due to lack of oil. The corresponding sliding surfaces of the rocker arms can be seen in figure 8-6. **Both the camshaft and the rocker arms can be repaired by engine reconditioning workshops.**

Camshaft with oil deficiency damage

Figure 8-4

Figure 8-5

Figure 8-6

8.1.5 Valves

Damage to valves can be caused by too much or too little valve clearance. When the sealing surfaces of the valves and the valve seats wear out, the valve clearance becomes smaller and the valve no longer closes properly. In this case, the valve seat "burns" and the hot combustion gases can flow through the valve, which doesn't close properly. A valve clearance, which is too large is initially less problematic, which is why the valve clearance is often set slightly higher than the predetermined value.

End face of the valve stem and corresponding area of the valve adjusting screw
Figure 8-7

A valve clearance, which is too large, is recognized by noise development, while a too small valve clearance is noticed only when damage has already occurred. However, there is a disadvantage if the valve clearance is to large. There is a low loss of power because the valve no longer opens wide enough. In addition the rocker arm with the adjusting screw strikes later and after a larger idle stroke and thus also faster on the end face of the valve stem. The adjusting screw then damages the end face of the valve stem (figure 8-7).

8.2 The clutch

In order to isolate the vibrations of the engine from the tooth flanks of the gear wheels, the torque of the engine is not transmitted by a rigid connection, but by means of six damping springs, which are arranged at the backside of the clutch housing.

This means that these springs are compressed with increasing torque – when accelerating - and they expand again when the torque decreases. Both the springs as well as their counterparts in the clutch housing are designed to be too weak for this permanent load alternation, so that the springs break and their counterparts in the clutch housing gets damaged. Figure 8-8 shows the the clutch housing with the gear wheel of the primary drive in a view from the rear (left). On the right, the riveted plate, which holds the damping springs, is removed. On figure 8-9, a broken damper spring and the damaged corresponding recess in the clutch housing are shown.

Rear of the clutch housing
Figure 8-8

Broken damper spring and damaged deepening for damper spring
Figure 8-09

Springs with a larger diameter of the wire may be a remedy, but they also provide less protection for the the tooth flanks of the gear wheels of the transmission.

Clutch boss with imprints of the clutch disks
Figure 8-10

Both the clutch friction plates and the pressure plates work with their edges into the respective grooves of the clutch housing and the clutch boss. Figures 8-10 and 8-11 show the imprints of the edges of the plates in the grooves of the clutch boss and the clutch housing. Figure 8-10 shows the clutch boss with the imprints of the edges of the clutch plates. Figure 8-11 shows an enlargement detail of the imprints in the grooves of the clutch housing.

Imprints of the clutch disks on the clutch housing

Figure 8-11

Rework is normally not necessary. Despite the imprints the clutch still separates the crankshaft from the transmission but only slightly slower because the edges of the clutch plates have to slip over the imprints. However, you should consciously shift the gears slower so that the clutch is disengaged during gear change. Otherwise damages to the transmission components will occur.

You should never try to smooth the imprints by hand with a file. It will not be possible to remove the same amount of material on all surfaces, so that one surface protrudes and initially carries the entire load alone. Here, the imprints will then re-appear much faster. If you want to smooth the imprints, you should only get this done in a workshop by the aid of a tool machine with a rotary table.

8.3 *Transmission*

Damage due to wear and misoperation can occur on the transmission and the shifting mechanism.

8.3.1 Gear wheels

In the case of the gear wheels, damage occurs to the tooth flanks, which can be influenced by the quality of the engine oil. Figure 8-12 shows the tooth flanks of a transmission with a mileage of approx. 160- to 170,000 miles with a strong pitting on the tooth flanks.

The pitting was found when replacing the ball bearing of the transmission output shaft, which had already a radial clearance – due to wear - of about 1 mm. The extent to which the defective bearing has favored the pitting is not known.

Pitting on the tooth flanks

Figure 8-12

Worn-out edges on the shift claws and the corresponding recesses, as shown in the next paragraph, are due to the fact that the clutch has been released too fast and thus the clutch was not completely disengaged during the shift procedure.

8.3.2 Shifting claws

The worn edges of the recesses in first speed gear wheel (Figure 8-13, left) and the worn edges of the shift claws of the fourth speed gear wheel (Figure 8-13, right) were the reason that the first speed suddenly disengaged when accelerating. The surfaces on which the shift claws still are engaged are reduced, so that such damage will progress rapidly. If wear is already visible on shift claws of gear wheels or on their counterpart, the recesses, they should no longer be used, since also the shift forks are affected.

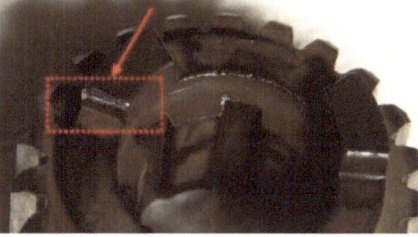

Gearwheel of the first gear on the transmission drive shaft

Gearwheel of the fourth gear on the transmission drive shaft

Figure 8-13

Since the shift claws slip out of the recesses - skip them - a force acts on the shift forks which moves the gear wheels in the axial direction of the transmission shafts.

Gearwheel of the fourth gear on the transmission drive shaft
Figure 8-14

The gear wheels, which are shifted in the axial direction on the transmission shafts by the shift forks, have circumferential grooves into which the shift forks engage as shown in figure 8-14. Figure 8-14 shows the gear wheel of the fourth speed on the transmission output shaft (the shift claws are shown in Figure 8-13). The force caused by the "skipping" of the shift claws has caused abrasion on the shift forks (Figure 8-15) and the circumferential grooves.

8.3.3 Shift forks

A similar material abrasion as at the circumferential grooves also arises at the ends of the shift forks, which engage in the circumferential grooves. The result is that the ends of the shift forks becomes narrower while the circumferential grooves get wider.

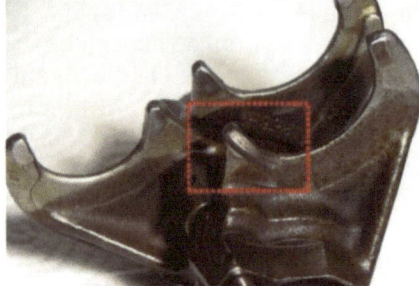 

Gear shift fork for shifting the first gear with material wear at the ends
Figure 8-15

The shift claws are thus no longer pushed far enough into the corresponding recesses of the mating gear wheel so that the area available for the force transmission becomes smaller. Thus the edges of the shift claws, as shown in figure 8-13 in an advanced state of wear, get rounded.

An condition of the shift fork for shifting the first speed as shown in figure 8-15 is the result. Shift forks that have an initial wear on their ends should not be used any more, even though it will for some time still be possible to ensure proper shifting. However, since the shift claws do no longer engage properly because they protrude no longer far enough into the rececces of the mating gear wheel, so that a sufficiently large area is available for the transmission of force, the damage as documented here will prematurely occur.

8.4 Oil circuit

8.4.1 Oilfilter

The filter sieve of the original oil filter, which is located at the lowest point of the engine housing below the crankshaft, tends to tear after only a short period of use, as shown in figure 8-16. It is possible to repair such oil filters by attaching a baffle and thus to prevent the filter material from tearing again.

Oil filter with torn filter sieve
Figure 8-16

However, such a repair is not advisable, since the filter material is too coarse to filter out fine metal particles. Fine metal particles, which damage the oil pump, are generated in operation e.g. due to normal wear of the teeth of the gear wheels. Internal paper filters, whose filter cartridges can be replaced, are available. The disadvantage is, that the cover under the engine housing has to be removed to replace the internal paper filter. A conversion to an external filter cartridge, as it is usual with passenger car engines, is also available. In this case, however, the ground clearance gets smaller.

The main disadvantage of a paper filter instead of the original sieve filter is its installation location on the suction side of the oil pump. A paper filter is usually installed on the pressure side of the oil pump, so that it tears before it gets clogged. By this way, oil is still supplied to the engine though it is not filtered. If the paper filter on the suction side of the oil pump is clogged, the oil pump can no longer suck in oil and a damage to the engine is the inevitable consequence. The installation of a paper filter instead of the sieve filter has the advantage that the oil is better filtered, but the disadvantage that in the event of a lack of maintenance a clogged filter can cause an oil deficiency which damages the engine.

8.4.2 Oil pump

Figures 8-17 and 8-18, show a heavily worn oil pump together with a drawing indicating the allowable wear dimensions.

The oil pump is worn out by metal abrasion, which was not filtered out by the oil filter. As the wear increases, the oil pump delivers less and less oil, which again leads to increased wear on other components of the engine and gearbox with even more abrasion.

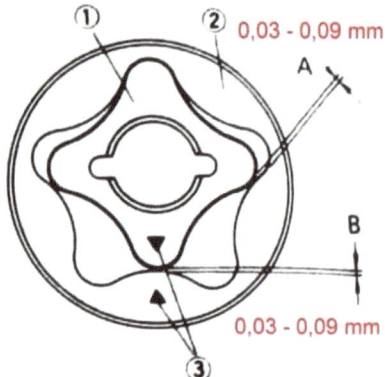

Outer rotor of the oil pump with wear dimensions

Figure 8-17

Cracked outer rotor of the oil pump

Figure 8-18

An external rotor of the oil pump with a crack caused by an unfiltered foreign body is shown in figure 8-18.

9 Electric system

The electric system of the XS 650 is relatively complex and the circuit diagrams in the workshop manuals may be confusing and difficult to read for those who are not trained car mechanics or electricians. There are many functions and components that you do not necessarily need. This makes the wiring harness complicated and the circuit diagrams difficult to read. Also, the susceptibility to failure increases, because where many cables are, the likelihood that one cable gets damaged and leads to a short circuit or leakage current is simply larger. A leakage current can discharge the battery while the vehicle is stationary, even if the circuit concerned is not actually in operation.

If there are problems with the electrical system and you suspect that the reason is a worn wiring harness, you should first simply replace it. Original wire harnesses are offered today for reasonable prices by spare parts dealers specializing in the Yamaha XS 650. If you only exchange individual cables, it will work again - but probably not long since all cables are the same old and worn. With a new wire harnesses the problem will be fixed for a longer time.

Another reason for problems can be oxidized plug connections, which have an excessively large transition resistance for the current. If the engine starts badly, first measure the voltage at the ignition coil. If a multimeter is used for measuring, as I will show it later, the absolute value of the voltage displayed is less important. Such instruments, which are bought for less than 10 $, are sufficient for this purpose, although they are often not very accurate. If you measure about 12.8 volts at the battery and 11.8 volts at the ignition coil, then this is o. k. One volt voltage drop between battery and ignition coil is perfectly normal. However, if the voltage measured at the ignition coil is 2 - 3 volts lower than the voltage measured at the battery, then this may be the reason why the engine starts poorly and tends to stall when idling. In any case, the reason for the voltage drop is a high transition resistance at the contacts in the ignition lock and in the plug-in connections. If the contacts have been cleaned, the voltage at the ignition coil will not be significantly lower than the voltage at the battery.

In order to correct a fault in the components of the electrical system, such as the generator, the voltage regulator or the rectifier, it is possible to replace these parts, if present, by tested parts. If the error is fixed, then you have found the defective part. However, such a search for an error is usually not possible, because not all the components are available for exchange. Then it is necessary to check the components which may not funktion properly. This is possible with simple means. However, you must be familiar with the function of the components.

9.1 Simplified electrical system

To make you familiar with the electrical system, I have described a simplified vehicle electrical system on the following pages, which I have installed in my own motorcycle. There are only the features that are required for driving on public roads. For the sake of clarity, I divided the circuit diagram into two areas, the "charging circuit" and the "consumption circuit" which are shown separately. In the " charging circuit " I kept the cable colors as in the original circuit diagram. In the "consumer circuit" I have chosen the color black according to the original for ground cables and the color brown for switched positive. In the wiring diagram as described here, only very few cables are necessary, so that it is possible to make a wire harness with fewer cable colors than I used it here. In order to be better able to describe the circuit diagrams, I have shown the circuit diagrams in color. So when I speak of a blue cable, a blue line in the circuit diagram is meant.

9.1.1 Consumer circuit

In order to keep the cables as short as possible, I installed the ignition lock under the right side cover. I have used the original ignition lock here, but it may be any other, as long as it has three ports with three key positions. It must be possible, that the key can be removed in position "one" and that there is no connection between the ports. In position "two" the ports "one" and "two" are connected and in position "three" the ports "one", "two" and "three".

Ignition switch close to the battery

Figure 9-1
The ignition lock is located on a holder below the right side cover. This makes the supply line from the battery and the connection to the fuse box in the right side cover as short as possible.

On the original ignition lock, a red cable (coming from the battery) is located at port 1, two brown cables (switched plus) at port 2, and a blue cable (head light and rear light) at port 3.

A fuse box (Conrad order no. 84 05 64-33, 7.95 €) is installed under the left side cover to prevent that a short circuit in one of the consumer circuits burns the only fuse in the original system. If the only fuse burns, this will mean, that the trip can only be continued after the problem is fixed.

Of the six existing fuses in the fuse box only five are required.

Figure 9-2
Behind the left side cover, there is a fuse box between the voltage regulator and the air filter with slots for 6 fuses. Conrad order no. 84 05 64-33

A brown cable from port 2 of the ignition lock is connected to fuses 2 to 6 in the fuse box. The blue cable from port 1 is connected to fuse 1. In position 1 of the key, all consumers and the ignition are supplied with power.

Headlight: Identification color: blue - Fuse 1

The blue cable from the ignition switch (port 3) is first linked to fuse 1 and then continued to the dimmer switch in the switch unit on the left side of the handlebar. From there a blue / green cable runs to the low beam light lamp and a blue / yellow one to the high beam lamp. Both lamps are connected via diodes (Conrad order no. 15 28 97-33, 0.56 €) to the parking light in the main headlight and to the instrument lighting. Another blue cable leads from the fuse 1 directly to the taillight.

Figure 9-3
The parking light bulb receives its current via two diodes (Conrad order no. 15 28 97-33), which connect them with the leads for high and low beam. The instrument light is connected to the terminal of the parking light (can not be seen in the photograph).

Ignition: Identification color: gray: Fuse 2
The gray cable runs from the fuse 2 directly to one cable of the two ignition coils. It doesn't matter wether to the grey or the orange cable. The other cable of the ignition coil is connected to the respective contact breaker behind the chrome cover on the left camshaft side. The contact breaker located directly on the base plate and marked "R" is responsible for the right ignition coil and for the right cylinder. The contact breaker on the auxiliary base plate marked ("L") is reponsible for the left side.

Brake light: Identification color: yellow: Fuse 3
A yellow cable connects the fuse 3 to the two brake light switches and then to the brake light in the tail light.

Turn signal lights: Identification color: gray / black, red, green: Fuse 4
From the fuse 4, a gray / black cable leads to the flasher relay and from there to the turn signal switch in the switch unit on the left side of the handlebar. From here, a gray / red cable and a gray / green cable lead to the turn signals on the right and left side of the vehicle. There are flasher relays with two and three ports. The original one has two ports as described here. When you buy a relay from a spare parts dealer, it may have three ports. Then one port is connected to the fuse, another one to the switch and a third one to the ground, Usually there are symbols, which indicate the purpose of the ports.

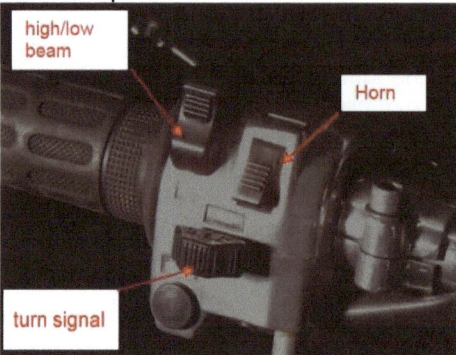

Figure 9-4

Horn: Color code: pink: Fuse 5
A pink cable connects the fuse 5 with the horn switch in the switch unit at the left end of the handlebar. From there it continues to the horn.

Remains free.: Fuse 6
I have connected fuse 6 to a voltmeter, which indicates if the ignition is switched on or off as well as the charge state of the battery.

The switch unit on the right side of the handlebar is no longer required. The switch in the left switch unit, originally intended for the headlight flasher, now operates the horn. A blind plug closes the hole for the original horn switch.

Additional volt meter

Figure 9-5

For safety reasons, it may also be advisable to secure the parking light separately to remain recognizable for others when the fuse or the bulb of the headlight is burnt. The blue cable from the ignition lock is then connected to the free fuse no. 6 and leads from there to the parking light bulb. It doesn' t matter when in addition a voltmeter is connected to fuse no. 6. Also an indicator lamp for the turn signals, which as well as the parking light bulb can be connected via two diodes to the left and right side, seems to me worthwhile, so that you do not forget to turn off the turn signal.

What I've described so far is the least what is needed. The installation should not cause any problems. It can be helpful if you imagine the cables are water pipes and the switches are stopcocks. Just as the water flows better through a thick pipe, the current also flows better through a thick cable. Everyone who has already installed a ceiling light in the house should also be able to install a simplified wire harness on a XS 650 motorcycle, as shown on figures 9-6 and 9-7 on the next two pages.

When installing the wire harness, lay the cables from the fuse in the fuse box to the switch and from there to the load (light bulb, horn etc). Then connect the load with a black cable to the frame or better directly to battery minus. Lastly, cover all cables with an adhesive ribbon. It does not necessarily look very professional, but is in any case not worse to the bougier pipe used in serial production.

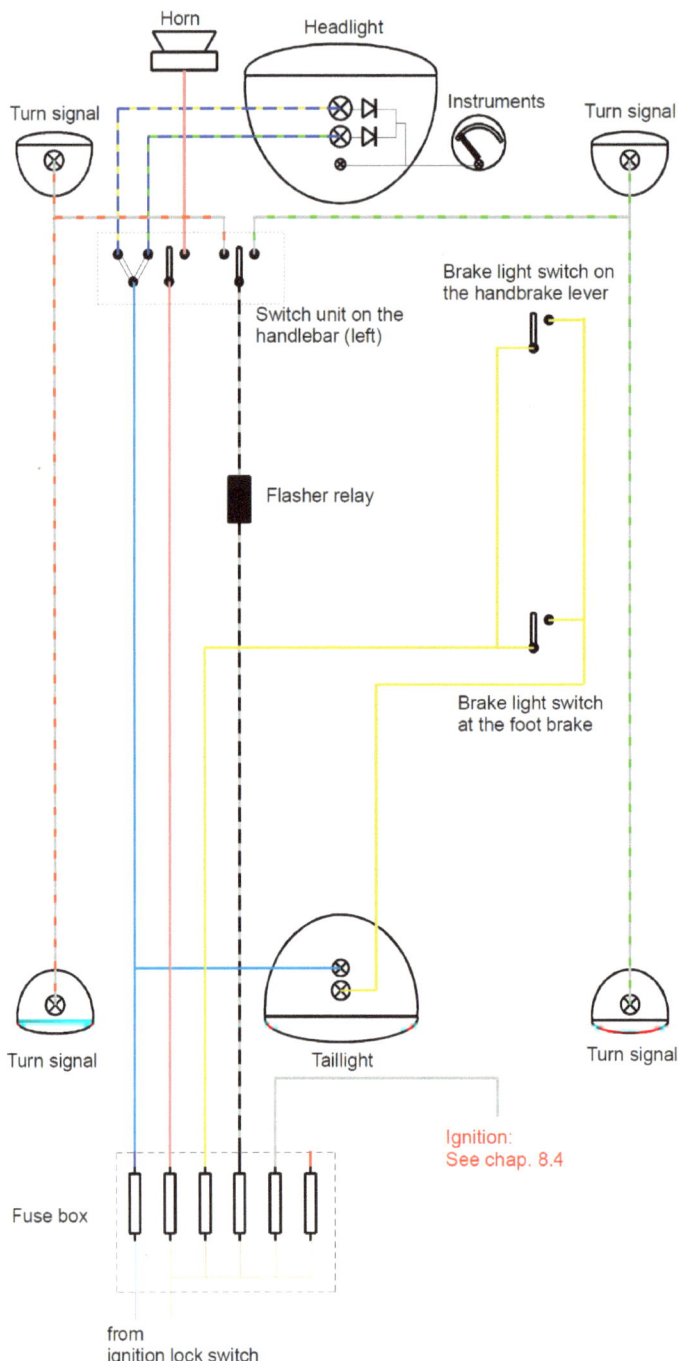

Figure 9-6

9.1.2 Charging circuit

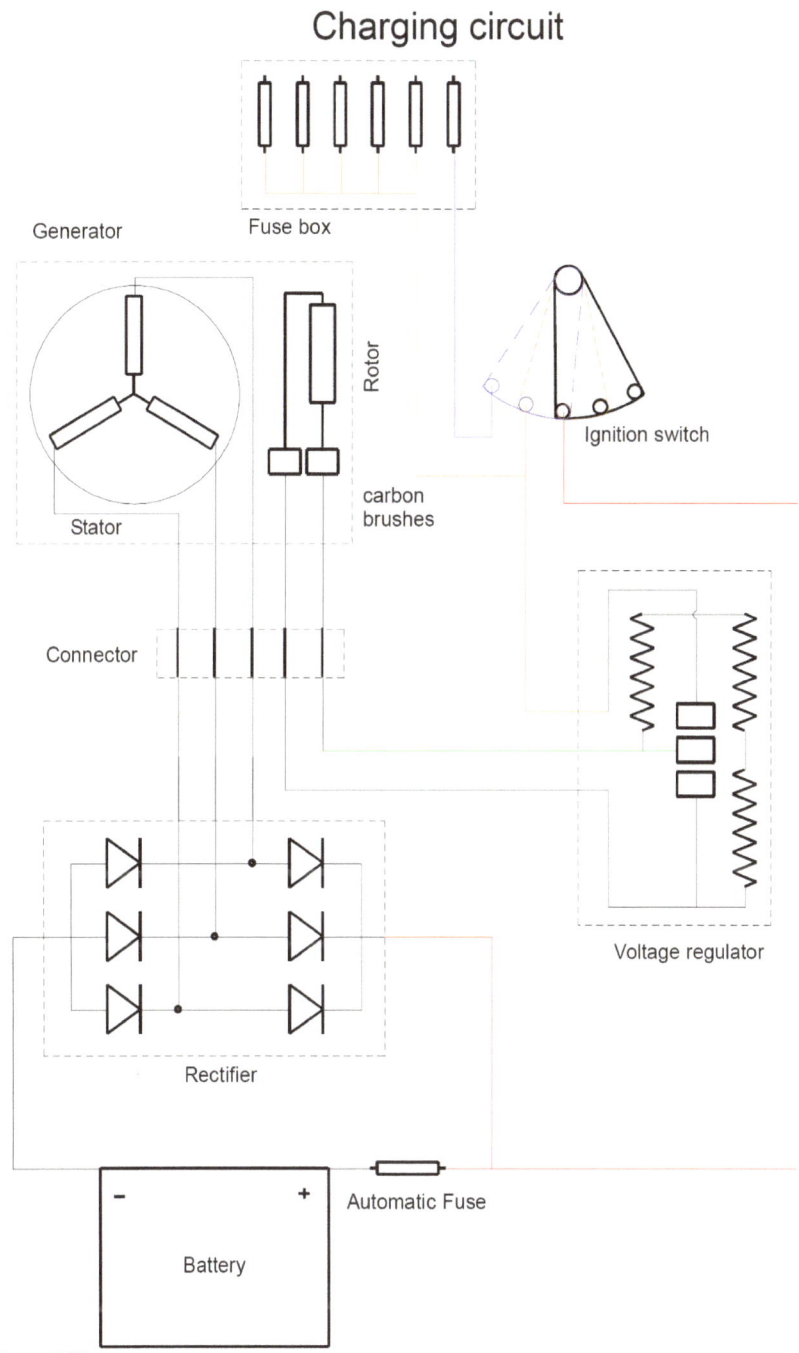

Figure 9-7

While the consumption circuit is still quite easy to understand, if one imagines that the cables are water pipes and the switches are stopcocks, this comparison is not enough to understand the charging circuit. Installing the wire harness according to the circuit diagram in figure 9-7 should not be a problem. However, in order to check the function of the components, one has to deal with this more closely.

9.2 Funktion of the charging circuit

9.2.1 The generator (alternator)

A "generator" is a device, which creates electric power. An "alternator" is a special kind of generator, which creates ac-current. So in the original manual the generator is sometimes refered to as the alternator.

First, a few basics:

If an electrical conductor is connected to a voltage (e.g. a battery) and brought into a magnetic field (e.g. a horseshoe magnet), it must be held with a certain force, otherwise it moves out of the magnetic field (the principle of the electric motor).

If an electrical conductor (e.g. a piece of copper wire) is moved with a certain force through a magnetic field (e.g. a horseshoe magnet), a current will flow (the principle of the "generator".)

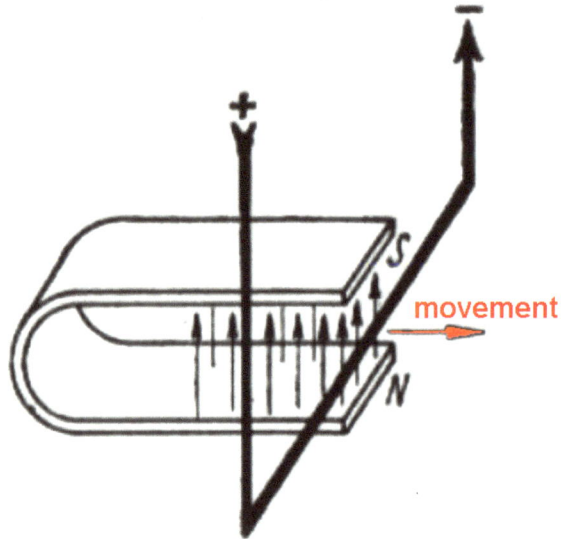

Figure 9-8

The direction of the current changes depending on whether the conductor is moving into or out of the magnetic field. Of course in a generator, not only a single conductor, but several loops - a "coil" - is moved through the magnetic field. In any case, an alternating current is always produced, as shown in figure 9-9, since the conductor loops are always moved into and out of the magnetic field.

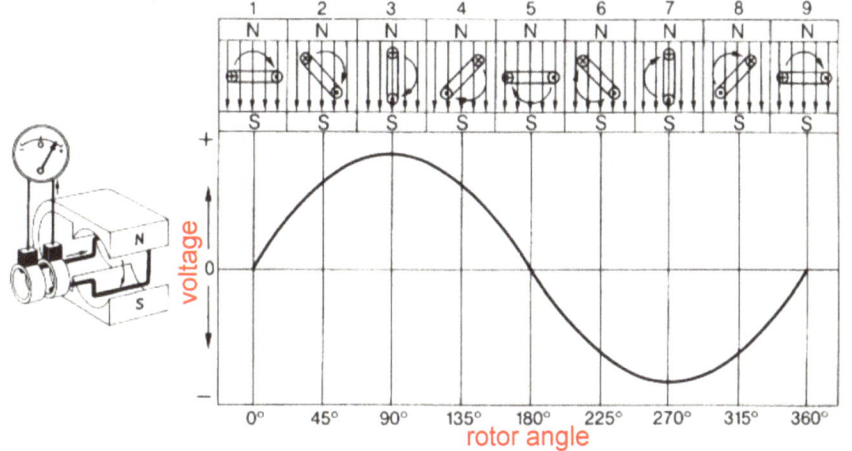

Figure 9-9

If not a single winding or loop is moved through the magnetic field, but three coils, which are offset by 120°, this is the operating principle of the three-phase generator, as it is used in the Yamaha XS 650. In the generator of the XS 650 the ends on one side of the three coils are connected and the ends on the other sides are led outwards (the three white cables coming out of the generator).

The generator, as it is installed on the left crankshaft stump, therefore consists of two components: the coils moving through a magnetic field, and the component that generates the magnetic field.

A very simple "generator" is a bicycle dynamo. Here, a permanent magnet rotates within a coil of copper wire. If you drive faster, the light is brighter, driving slower, it becomes darker. But this kind of behavior is not suitable for a motorcycle. The light has to be of the same brightness, no matter whether you are driving fast or slow, you have to be seen by the other drivers. This is why a magnetic field with a controllable strength is needed.

The generator of the XS 650 is in principle similar to the bicycle dynamo. The component, which generates the magnetic field rotates with the crankshaft, while the component, in which the current is created due to the magnetic field is bolted to the engine housing. In order to supply a current to the rotating component, a further component is necessary, the carbon brushes.

The following figure shows the generator of the XS 650 with the left engine cover removed.

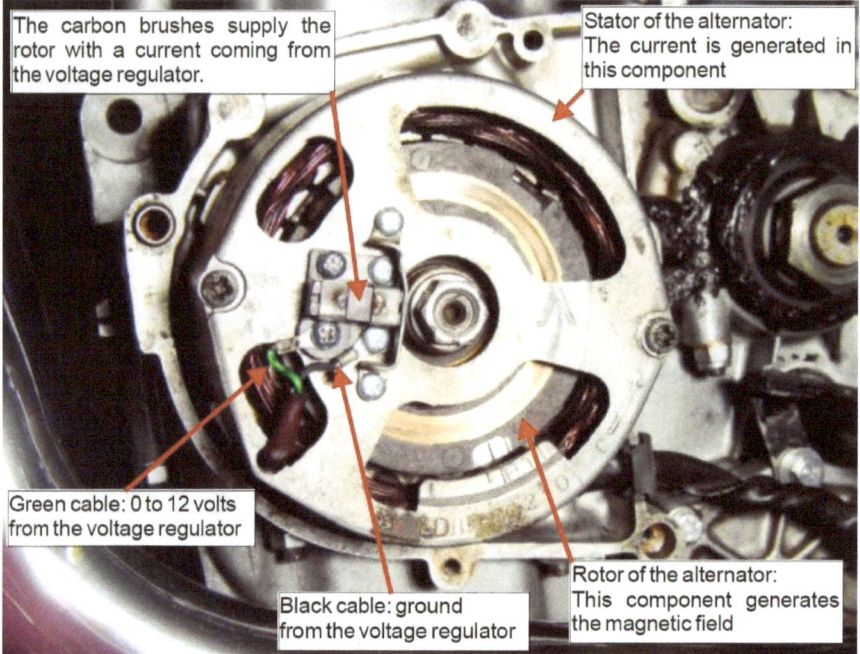

Figure 9-10

In workshop manuals the component of a generator, which is connected to the engine housing and which is stationary in operation, is refered to as the "stator".

The component which rotates is refered to as the "rotor".

From these designations, it is not yet clear which component has which function. Although I do not consider these terms to be very apt, I will use them in the following descriptions to not confuse those who have already read in the original workshop manual on the topic of electrics.

The stator
On the left side of Figure 9-11, the circuit symbol for the stator is shown, as used in the circuit diagrams in the workshop manual. The three rectangles arranged at an angle of 120° symbolize the three coils which produce the three-phase current. On one side, the coils are each connected, while the other side is goes to the outside. These are the three white cables that are found in the connector to the engine.

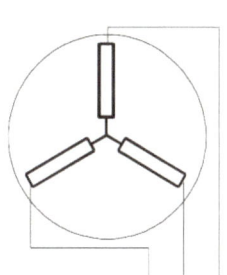

Figure 9-11

Furthermore, the carbon brushes, which connect the regulator to the rotor, are also located in the housing of the stator. With the stator they have nothing else to do.

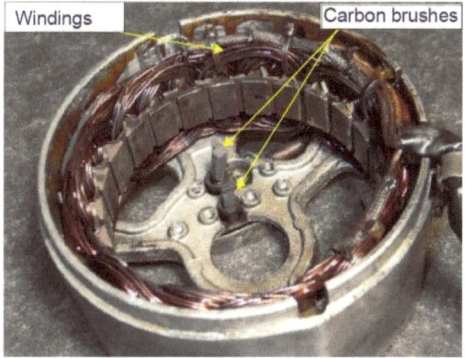

Figure 9-12

Since the stator does'nt move, mechanical damage, except perhaps by vibrations, is rare. If a stator has to be replaced, the usual reason is, that one or more windings are "burned". Everybody can imagine, what is meant by a "burnt fuse". The thin wire of a bladetype fuse has become glowing, and the liquid metal of the wire has dripped down. The connection is thus interrupted.

In the case of a winding, the term "burnt out" means something different. In a single winding only a very low voltage is induced. In order to obtain a sufficiently high voltage, several windings have to be moved through the magnetic field. To arrange the windings as closely as possible, an insulating lacquer is used to insulate the windings which would be otherwise in contact with each other. If the windings become too hot, the insulating lacquer can melt and the windings will get in direct contact with each other. When two windings are in contact, the current no longer flows through both windings but it takes the shorter path through the point where the windings are in contact. This means, that one winding is lost for creating voltage. In addition, the electrical resistance of the coil is thus reduced because the current can travel a shorter distance from one end to the other. The coil is not immediately useless, but it decreases in performance. The performance can be checked by measuring the electrical resistance of the coil.

The wire of the windings of the stator has a diameter of approximately 1 mm including the insulating lacquer. This corresponds to a cross-section of approximately 0.75 mm². According to the workshop manual, the electrical resistance measured between two white cables should be about 0.8 to 1 ohm. For a resistivity of copper of 0.0185 ohm mm² / m, 1 ohm corresponds to a line length of approximately 40 m, which the current must pass from one end to the other. If a lower resistance is measured here, this means that the current has to pass a shorter distance. For more information, see chapter "9.3 Check the components of the charging circuit"

The rotor
The rotor consists of a coil whose ends are accessible via the slip rings. The circuit symbol of the rotor is a rectangle, which stands for resistance in electrical engineering.

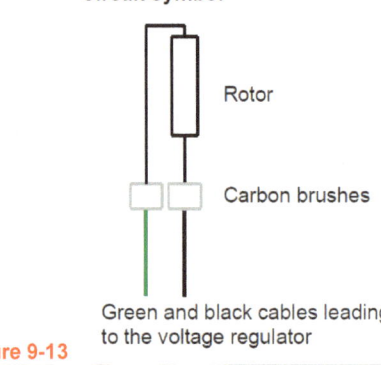

Figure 9-13

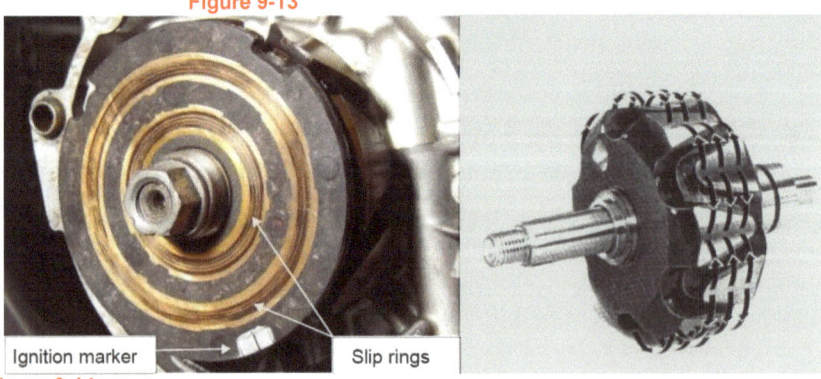

Figure 9-14
On this similar rotor, the course of the field lines of the magnetic field is represented by arrows. It can be seen that the arrows point in different directions.

The coil of the rotor is made from copper wire, which is provided with an insulating lacquer, like that of the stator. For the "burn out", the same applies as described in the chapter "stator".

Such a coil has of course not only an electrical resistance, which is measurable with an ohmmeter. It has also an inductance – that means, it generates a magnetic field. To explain the term inductance would excede the topic of this book. The correct curcuit symbol for a coil, which always has also an ohmic resistance, is a filled rectangle. I have used an unfilled rectangle for both the stator coil and the rotor coil to use the same symbols as in the original manual. On the end face of the rotor there is the ignition mark, which is not concerned with function of the rotor as a component of the generator.

The carbon brushes

The carbon brushes are made from graphite and have a length of 15 mm and a cross-section of 4.5 x 5 mm with an inlet copper cable. The task of the carbon brushes is to transfer a current to the rotating rotor, which generates a magnetic field. The figure below shows a carbon brush in a new condition, an already used one with a soldered adaptor and a worn out carbon brush.

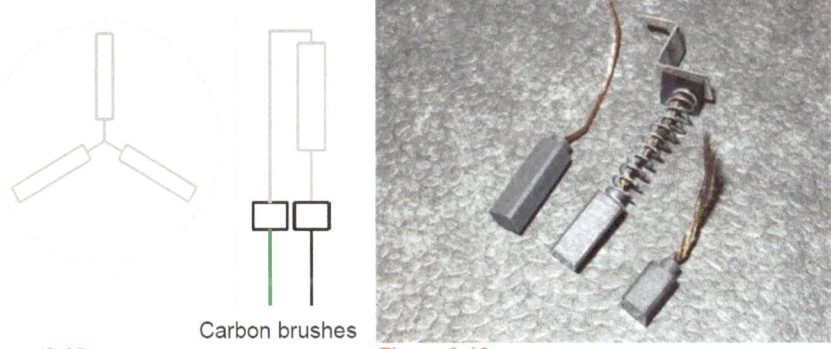

Carbon brushes

Figure 9-15 **Figure 9-16**

You can buy the complete carbon brushes from Yamaha or re-use the adaptor and the spring. The graphite part with the inserted wire can be procured considerably cheaper from part dealers. Carbon brushes are wear parts, which after about 10,000 km are worn, i.e. they have got so short that the pressing force of the spring is no longer sufficient to ensure a reliable current flow to the slip rings of the rotor.

9.2.2 The rectifier

As described above, the generator, which consists of the rotor and the stator can only supply alternating current which constantly changes its direction. With alternating current, however, you can not charge a battery. The alternating current must be rectified so that it flows only in one direction. Just as there are non-return valves for fluids that let the fluid pass only in one direction, there are diodes for the electric current that allow the current only to flow in one direction. Since the alternator has three coils, which are led out with the three white cables, three currents must be rectified.

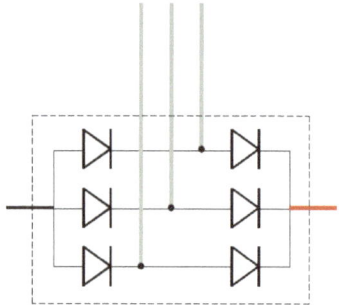

Figure 9-17

In figure 9-17 above, you can see the circuit symbol of the rectifier, as shown in the wiring diagrams. Figure 9-18 shows the original rectifier, as it is installed underneath the battery. Figure 9-19 shows an inexpensive alternative of Conrad Electronics – alone and with an attached heat sink, as it is also to be purchased from Conrad Electronics or others.

Figure 9-18

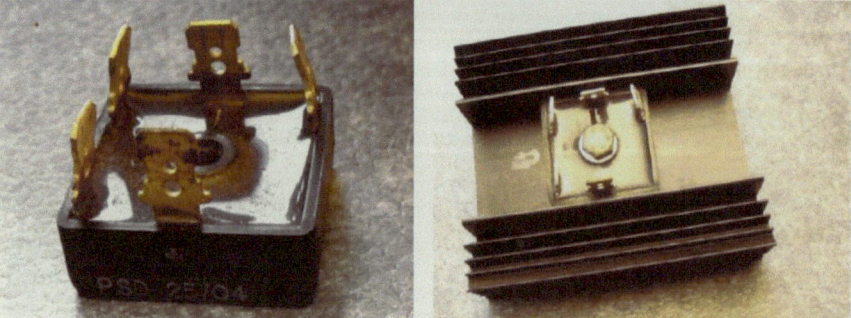

Figure 9-19

The circuit symbol oft the rectifier explains the function quite good. To illustrate it more, I have built a rectifier using six diodes, as I have them also used for the parking light bulb in the main headlamp (Chapter 9.1.1 Consumer circuit).

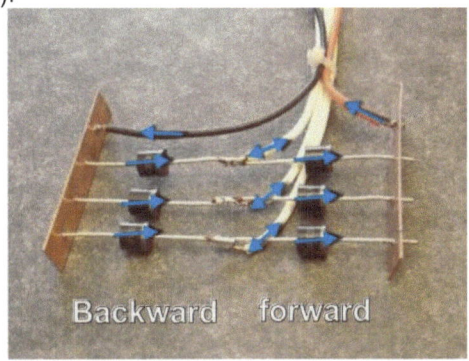

Figure 9-20

Such a rectifier is still a lot cheaper than that purchased from Conrad Elektronik (Fig. 9-19) and electrically fully functional. Whether it is also able to cope with the mechanical loads in driving operation is not sure, but in an emergency it surely will work for some time.

The picture shows the three white cables coming from the generator, the red one which leads to the ignition lock and the battery positive pole, and the black one coming from the battery negativ pole. The currents in the three white cabels from the coils of the stator change more or less quickly their direction, depending on the speed of the engine. They are either passed by the diodes in the "forward" branch into the red cable, or are blocked by the diodes in the "backward" branch. The currents then flow through the red cable to the consumers and to the battery and back from there through the black cable and the diodes in the "backward" branch to the coils in the stator, thus the circuit is closed.

9.2.3 The voltage regulator

Just as the rotational speed of the engine not only changes the direction of the current in the coils of the stator more or less quickly, also the voltage increases with a higher engine speed. For charging the battery, however, a constant voltage is required which is slightly above the rated voltage of the battery of 12 volts. The headlights also require a quite constant voltage of 12 volts.

To remind you once again of the function of the generator:
A current flows through the coils of the rotor, which generates a magnetic field. Inside the rotating magnetic field are the coils of the stator, into which

a voltage changing with the engine speed is induced. The amount of this voltage depends not only on the engine speed but also on the strength of the magnetic field generated by the current flowing through the rotor. The current flowing through the rotor and thus the strength of the magnetic field can easily be influenced by the voltage that is applied to the carbon brushes.

The purpose of the voltage regulator is to adjust the voltage, that is applied to the carbon brushes, so that the voltage induced into the stator coils is about 14 volts. This is the voltage required to charge a 12 volt battery.

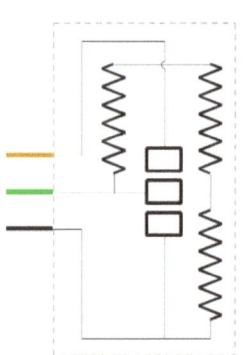

Figure 9-21   Figure 9-22

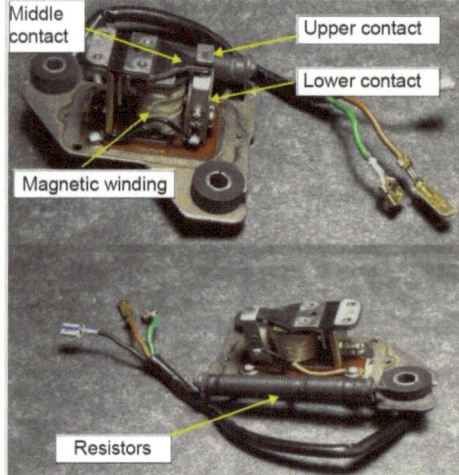
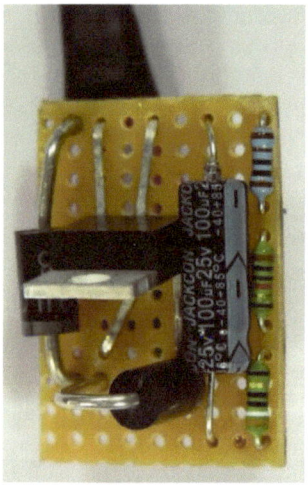

Figure 9-23   Figure 9-23a

Figure 9-21 shows the voltage regulator as a circuit symbol. Figure 9-22 and 9-23 show the voltage regulator as it is installed under the left side cover and with its cover taken off. As a comparison figure 9-23a shows a selfmade electronic voltage regulator, which works by the principle of pulse width modulation. It is much smaler than the original one, more reliable and there are no resistors, which act as a consumer themselves. Electronic voltage regulators are offered by parts dealers for about 30 $.

To the brown cable, the board voltage is applied. If the board voltage is too low e.g. when the engine is stopped, then only the actual battery voltage is applied. The green cable thus supplies the highest voltage available in the on-board network at this time. If the engine is now started, and if the engine speed increases, the voltage that is applied to the green cable must be reduced, otherwise the bord voltage will become too high. This would overstress the generator, the battery can start to boil and bulbs can burn through.

In order to reduce the bord voltage, which is applied to the green cable and thus generates the magnetic field, which induces the voltage, the green cable is no longer directly connected to the electrical system, but resistances are interposed.

As shown in the circuit symbol (figure 9-21), the brown cable is connected to the upper contact, the green cable is connected to the middle contact, and the black cable (ground) is connected to the lower contact. In addition, the brown cable is connected to the coil of the magnet. If the bord voltage is low, the middle contact is pressed against the upper contact by a spring and there is a direct connection between the brown and the green cable.

As the bord voltage increases, the magnetic field also becomes stronger and the magnet pulls the middle contact away from the upper contact. As long as the middle contact is between the other two contacts, the current flows from the brown cable through a resistor and through the green cable to the rotor coil. Because the current must first pass the resistor, the voltage at the rotor coil is, of course, lower, which causes a weaker magnetic field around the rotor.

If the bord voltage continues to rise, the magnet in the voltage regulator pulls the middle contact, which is connected to the green cable, against the lower contact, which is connected to the black cable. Now both ends of the rotor coil are connected with ground (negative) and there is no longer a magnetic field. The bord voltage drops now and thus the voltage in the magnet coil in the voltage regulator drops too. The spring can now pull the middle contact back from the bottom contact so that a current can flow from the brown cable via the resistor to the green cable. Now a magnetic field builds up again around the rotor and a voltage is induced again into the stator coils.

If, however, the vehicle board voltage continues to drop because, for example, the engine speed is reduced or a consumer is connected, the force of the magnet is reduced and the spring moves the middle contact against the upper contact. Now the highest available voltage at this moment is again applied to the rotor coil via the green cable. This procedure is repeated within fractions of seconds.

It becomes clear that the force of the magnet in the voltage regulator, the force of the spring and the resistance must fit together. The pretension of the spring can be adjusted with the adjusting screw (Figure 9-24). Since the current flows through the coil of the magnet and through the resistor, the coil and the resistor become warm during operation. For conversions, you should always make sure that the voltage regulator - as it applies to all electrical components - gets enough cooling air.

Figure 9-24

9.2.4 The ignition lock

The ignition lock should be connected to the battery by a cable as short as possible. This ensures on one hand that the vehicle can not be used unauthorized, on the other hand it separates the entire electric system from the battery when the vehicle is not used. If you forget to turn off the light, the battery will be empty after some time. If one forgets to switch off the ignition, a current flows through one of the two ignition coils. The current not only discharges the battery, but it heats up the ignition coil, which can cause windings to burn through.

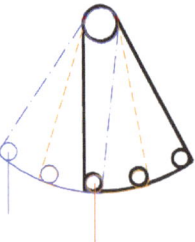

Figure 9-25 (circuit symbol)

If the wiring harness becomes older, the insulation is usually brittle and can cause leakage currents due to moisture which will discharge the battery even when the vehicle is stationary. Such leakage currents should not be able to flow, at least when the vehicle is stationary. This can be achived by separating the battery from the complete electronic system with an ignition lock very close to the battery. I therefore installed the ignition switch directly underneath the battery and thus turned off a potential location for leakage currents - the cables in the area of the steering head.

A simple key switch can be used as the ignition lock, in which the key can be removed in the switched-off state. But it is advantageous if the ignition lock has at least three positions: in the first position, the battery disconnects from the board electrical system and the ignition key can be removed. In the second position, all consumers and the voltage regulator except the headlight – as long as this is allowed in your country - are connected to the battery. So you can start the engine. After the engine has started, you can switch on the head light by turning the key into the third position.

There are three cables on the ignition lock, a red one as an input from the battery, a brown one for switched plus (position 2) and one blue for the head light (position 3). In position 3, the red cable is connected to the brown cable and blue cable. In position 2 the red cable is only connected to the brown cable. The original ignition lock has a 4th position with a 4th cable, which is not used here.

9.3 Check the components of the charging circuit

Electrical malfunctions often manifest themselves in the way that the battery is often "empty". There are several reasons for this. The battery is already quite old and needs to be replaced. If you turn on the headlight when the engine is not running and the light becomes noticeably darker after one or two minutes, you should buy a new battery.

When the battery is o.k. there are two reasons why the battery may be often "empty":

There are too many consumers that are discharging the battery, that means, a larger current is drawn from the battery than the generator is able to supply.

According to the specifications issued by Yamaha, the generator provides a current of 11 amps with a voltage of 14 volts at min. 2000 rpm of the engine. 14 volts are necessary, because the charge voltage must be slightly higher than the rated voltage of the battery. With the engine idling, the generator supplies of course less than 11 amps and the voltage is lower than 14 volts.

Through a 55/60 watt H4 bulb in the main headlights flows a current of:

55 watts / 12 volts = 4.6 amps.

In addition there are also the taillight, the ignition, the turn signal lights with 2 x 21 watt (3.5 amp) and the brake light with further 21 watt (1,75 amp). So when the battery of a motorbike, which is mostly rode in the city with stop and go traffic, is frequently "empty", this is not a defect. It is because the generator of the XS 650 was not designed for the requirements of today's city traffic.

In the seventies, there was not yet the rule to drive in daylight with low beam switched on. H4 light was not available as standard, but bilux lamps with max. 45 watt for low beam and high beam.

With old wire harnesses there is the tendency that the insulation gets brittle. Particularly in the case of wet conditions, very small leakage currents can flow, which do not "burn" the fuse, but discharge the battery over time.

Keep in mind, that the battery of the XS 650 with a capacity of 14 Ah (original) is much smaller than a car battery. If you don't use the electric starter, you may have installed an even smaller battery. But no matter what the capacity of the battery is, it has 12 volts and the leakage current, which leaks from a brittle cable is the same no matter if the battery has a high or low capacity. So a leakage current may not be recognizable in a car with its big battery, but the same leakage current will make the small battery of a XS 650 motorcycle frequently empty. Since leakage currents are not a special feature of Yamaha XS 650, I won't explain here in much detail how to find and eliminate them.

If one can exclude the reasons mentioned so far for a frequently "empty battery", one should look for the fault among the components of the "charging circuit". In order to be able to check the components properly, one should be familiar with their function.

In the search for faulty components one can proceed systematically and is then sure, if everything was tested correctly, that the malfunction is not in the power supply. The search for leak currents requires more patience and intuition since the leak current does not always have to occur, but e.g. only in the case of a certain position of the handlebar or in the rain.

9.3.1 Measuring instruments

In order to check the electrical system of a vehicle for faults you need measuring instruments. In former times, special measuring devices for the car sector could be purchased from mail order companies and from auto accessories shops. Nowadays the electrical systems of cars are far more complicated and less susceptible to faults, therefore these devices are hardly offered today.

Today, there are quite a lot of cost-effective digital multimeters available, with a lot of features we do not need for the vehicle electrical system. On the other hand they cannot do everything, that is needed for maintenance work on car electrics. However, the multimeters are very cheap and this can be accepted.

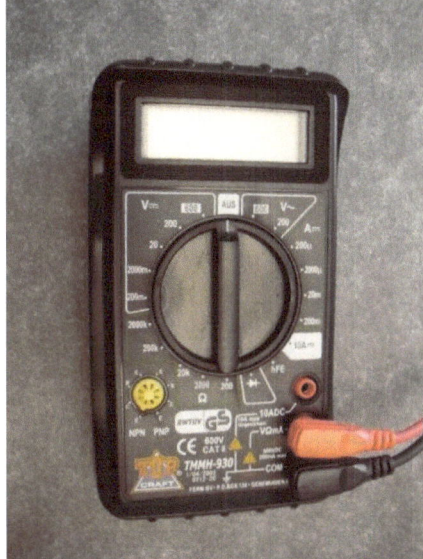
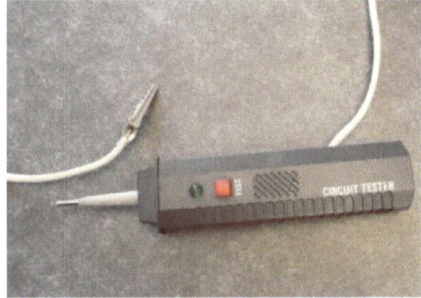

Figure 9-26

I bought the multimeter in figure 9-26 on the left for 3.95 € at Aldi and it has in principle everything we need except a stroboscope gauge and a dwell meter.

Such a multimeter should in any case be capable of measuring voltages up to 20 volts accurately (± 1 volt). That means, it should have a measuring range of 0 to 20 volts. In order to measure the resistances of coils in the rotor and stator as well as the resistance in the voltage regulator, an ohm measurement range is required that is as small as possible.

The nominal value of a stator coil is 0.8 to 1 ohms. With a measuring range of 0 to 200 ohms, as that of the multimeter from Aldi, such a measurement is quite inaccurate, but more expensive multimeters do not have a smaller measuring range either. However, to detect leakage currents, this is not sufficient. The device in figure 9-26 (continuity tester) on the right in the simplest version is available in hardware stores for about 10 €. It has a built-in battery and you can check cables whether they are broken. If something between the probe and the crocodile clamp lets a current from the built-in battery pass, a buzzer sounds and the green LED lights up when the red button is pressed. The device is very sensitive. If the battery of the motorcycle is disconnected, no passage should be detected between the cables of the battery, when all the consumers are switched off.

9.3.2 Check for leakege currents

There are many tips on how to detect leakage currents that are slowly discharging the battery when the vehicle is stationary. I do not want to explain this too much, because this is a description of the components of the electrical system and a guide for checking the components of the charging circuit. A description of all possibilities which can lead to leakage currents would be too extensive here. In order to be successful in the search, one should simply keep in mind, that a leakage current is a current which, while bypassing the actual consumer, flows from the battery positive pole or from the fuse box or a switch to the ground and which is so small that the fuse does not respond. In this circuit, a "passage" must be ascertained when the load is switched off, or a resistance which is less than "infinite". At infinity, the reading of a digital multimeter is usually "1" or the pointer of an analog measuring device remains at the right end of the scale. The smaller the resistance, the larger the current that can flow here and discharge the battery.

To find such a "leak", you can first replace the only fuse or the fuses of the individual consumers with a light bulb (starting with 21 watts). If it lights when the consumer is switched off, this circuit contains a leak where a small current can flow through the leak to the ground. The search becomes more effective if the consumers have individual fuses. Next, remove the battery and connect an ohmmeter or the device in figure 9-26 instead of the battery. In this case you should also move the handlebars slowly, since cables in the area of the steering head tend to fraying. It may be, that the leakage current only occurs at a certain angle of the handle bar.

9.3.3 Checking the charging current

As already mentioned in chapter "9.2 Function of the components of the charging circuit", the alternator delivers ~14,5 volts at 2000 rpm, as specified in the workshop manual. At lower engine speeds, e.g. when idling at 1200 to 1500 rpm, it delivers less. Therefore, the head light is darker when idling than at engine speeds higher than 2000 rpm. The simplest method to test the generator is therefore to light a wall with the head light and to open the throttle a little bit. The light must get brighter now. If not, there is something wrong in the charging circuit and one must expect that the battery will get empty during the next trip. It can be easily observed in "stop and go" traffic in the city, where more electric power is usually consumed than the generator can deliver. The own head light radiates to the rear of a standing car. When you start, the engine speed gets higher than 2000 rpm and the head light must become brighter. Even more effective is, of course, a voltmeter in the cockpit, where you can always check whether the battery gets charged or not. But not everybody likes a bike with a lot of instruments.

When you have read chapter "9.2 Function of the components of the charging circuit" before, you will know, that the generator can only produce electric power, when a voltage is applied to the rotor. When the engine is stopped, the electric power is delivered by the battery. The nominal voltage of the battery is high enough that it energizes the rotor at low engine speeds.

First, you should check whether this voltage is also applied to the carbon brushes. In the standard wire harness the current must flow from the battery through the steering head to the ignition lock, the voltage regulator and then to the rotor. On this way a few tenths of a volt can be lost - but it should not be more. So first measure the voltage directly at the battery and then at the carbon brushes as shown in figure 9-27.

Figure 9-27

The difference between the voltage measured at the battery and the voltage measured at the carbon brushes should not be more than 0.5 volts. Now start the engine. Up to an engine speed of approx. 2000 rpm the battery voltage should be measured at the carbon brushes. If the voltage at the carbon brushes drops earlier, the battery is only charged at higher engine speeds. This may be no major problem for those who ride mostly at higher speeds on country roads and highways. If the voltage drops only at higher engine speeds than 2000 rpm, the generator is overstressed and battery may start to cook. Here, the interaction between the spring and the magnet in the voltage regulator does not work properly. See chapter "9.3.8 Testing the voltage regulator" for more information.

9.3.4 Testing the rotor

The rotor is the component in the charging circuit, which is the most likely to be the source of a malfunction. In principle, the rotor is very easy to check.

Rotor

Figure 9-28   Figure 9-29

The rotor consists of a coil whose resistance is ~ 5 ohm. When the coil wears out, e.g. by "burning" of windings, the resistance can only become smaller since then the insulation between two windings is "burned", and the current can flow though a shorter distance from one slip ring to the other. If a higher resistance than 5 ohm is measured, this is in any case a measurement error. The cause may be that the slip rings are oxidized and thus represent a resistance which is, of course, also measured in addition to the resistance of the coil. Therefore, you should thoroughly clean the slip rings before measuring. For this purpose, abrasive papers with a grain size of 600 or fine steel wool can be used.

For measuring, the stator has to be removed. A current can flow via the carbon brushes in the event of a defect in the voltage regulator or in case of a leakage current in the cables between the carbon brushes and the voltage regulator. This may look like a too low resistance of the rotor. The ohm meter lets a current flow through the resistor to be measured. The current, which, to put it simply, comes back, is a measure of the resistance.If you are now sure that everything has been done correctly in the measurement, and if a resistance of less than 5 ohms is established, or if a passage from one of the slip rings to the ground is established, the rotor should be replaced. Between a slip ring and the ground, the ohm meter must indicate an infinitely large value. On a gauge with a pointer, the pointer is at the right end of the scale. A digital gauge doesn't indicate anything. Replacement for defective rotors is available at XS650 parts dealers.

A lower resistance than 5 ohms means that the insulation between some windings is burnt through. The remaining windings, of course, can no longer build up such a strong magnetic field in the generator as it would be necessary for the required power. It is compensated for by the fact that the excitation current flows through the remaining windings for an extended period of time. As a result the remaining windings get warmer. The insulation starts to melt (burnt windings) and even fewer windings are available. The damage is therefore accelerated and the rotor should soon be replaced.

9.3.5 Testing the stator

Figure 9-30 shows the stator with attached carbon brushes. The carbon brushes represent the connection from the voltage regulator to the rotor and are not a component of the stator. The circuit symbol on the right shows that the three windings of the stator are connected in the middle. The other ends are accessible via the three white cables.

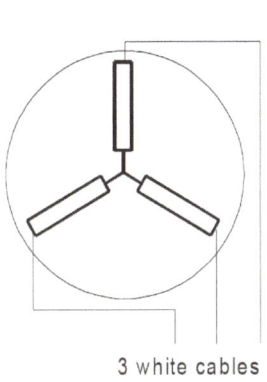

Figure 9-30

To test the functionality of the stator, the resistance between two white cables is measured. In total of three measurements are necessary. The ohm meter must display 0.8 to 1.0 ohms. The yellow cable (not shown here) must be disconnected. Otherwise you will not only measure the resistance of the coils. It is clear that such a measurement with a measuring range of 0 to 200 Ohm, which is the smallest measuring range of most multimeters, is not really precise. Finally, it is checked whether there is a passage for a current between the individual white cables and the ground. How to do this, I described on the previous pages when testing the rotor.

9.3.6 Testing the carbon brushes

The task of the carbon brushes is to transfer the current from the voltage regulator to the rotating rotor. One therefore disconnects the voltage regulator - so one is sure that the current from the multimeter set to resistance measurement can really only flow through the carbon brushes. One probe of the multimeter is placed on a slip ring and the other on the end of the green or black cable. The outer slip ring belongs to the green cable (+ from the voltage regulator) and the inner slip ring belongs to the black cable (ground). Figure 9-31 shows the circuit symbol, a new carbon brush, a still usesable one with a soldered adaptor and a spring, as well as a worn out carbon brush.

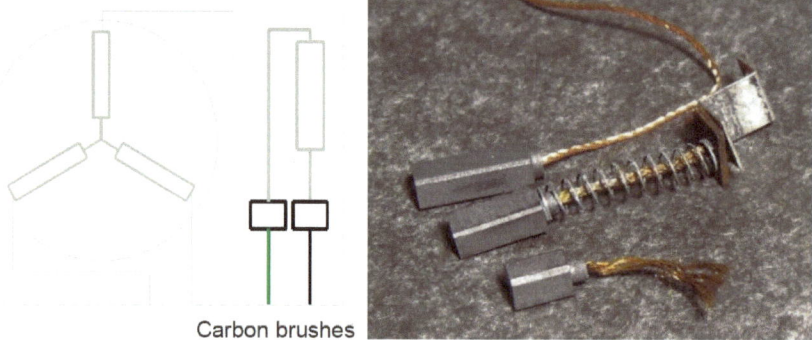

Carbon brushes

Figure 9-31

The ohmmeter should now display a resistance which is as small as possible. If the ohmmeter doesn't show anything at all, this means, that there is no contact between the carbon brush and the slip ring. In this case it can be due to the fact that the carbon brush clamps in its guide and it must be made moveble again. It also may be, that it is so worn that it no longer touches the slip ring. Then the corresponding carbon brush has to be replaced. It may also be, that you can read a resistance of 1 or more ohms on the ohmmeter. Then the slip rings must be cleaned with abrasive paper with a grain size of 600 or fine steel wool.

9.3.7 Testing the rectifier

The rectifier is very easy to check. The best way is to do this with a continuity tester, since the current flows from the built-in battery back to the device via the component to be tested and the alligator clip. The ohm meter shows a numerical value which is very close to zero in the forward direction of the diode. In the blocking direction of the diode nothing should be displayed. Be sure that the ohmmeter is connected in the correct direction.

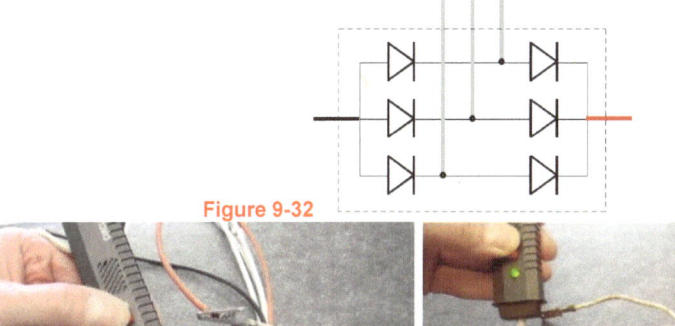

Figure 9-32

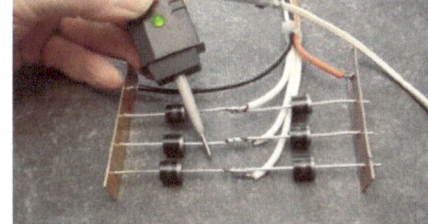 

Figure 9-33

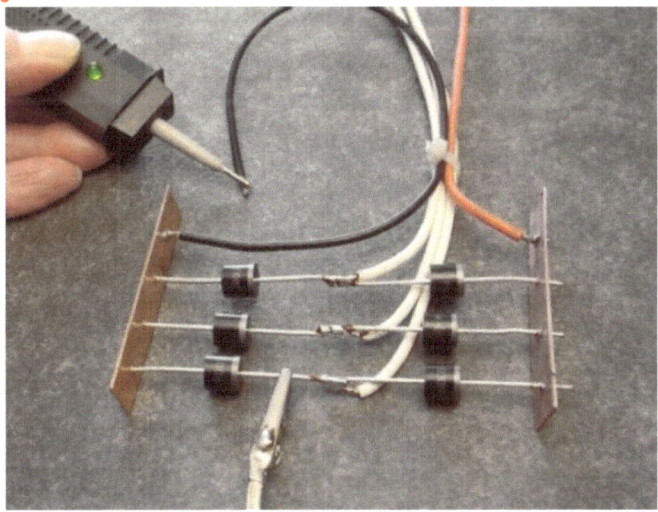

Figure 9-34

Figure 9-33 shows the test with the original rectifier. To make the whole thing more vivid, I demonstrate the test with a model (Fig. 9-34).

The current flows through the three white cables coming from the stator of the generator, then through the diodes and through the red wire into the vehicle electrical system (battery). It is therefore necessary to check whether there is a passage **to** the red cable from each white cable. Then, the continuity tester is turned. There must be no passage **from** the red cable to any white cable. In order to close the circuit, the current must be able to flow back into the stator coils.

To test this, put the probe of the continuity tester on the black cable. All white cables are checked for continuity. Then the test is repeated in the other direction. Now the continuity tester must not respond. If it has been determined that no current flows through a diode in the intended direction, or that a current can flow in the opposite direction, the generator cannot supply the full electrical power to the electrical system. The rectifier should be replaced then.

9.3.8 Testing the voltage regulator

To test the voltage regulator, we need an ohm meter. The voltage regulator consists of a changeover switch (three positions), resistors through which the excitation current flows to the rotor as the vehicle voltage increases, and a magnetic coil. Measurements are taken in the three positions of the changeover switch.

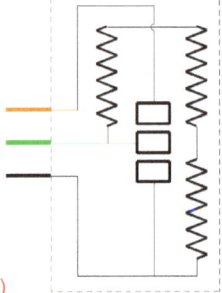

Figure 9-35 (circuit symbol)

Before the measurement, the contacts of the switch should be carefully cleaned as a transition resistor on the contacts distorts the measuring result and, of course, also affects the function of the voltage regulator. For cleaning, abrasive paper with a grain size of 600 can be used

Middle contact in upper position
In this position the coil of the magnet is checked and the passage from the brown cable to the green cable from the electrical system to the rotor.

Figure 9-36

The magnet coil should have a resistance of 36 to 38 ohms - measured between the brown and the black cable. Between the brown and the green cable, the resistance should be as small as possible, i.e. 0 ohms

Middle contact in middle position
In this position the resistor is tested through which the current from the vehicle electrical system flows when - with increasing bord voltage - the magnet develops so much force that it pulls the middle contact away from the upper contact.

Figure 9-37

A resistance of 10.7 Ohm should now be measured between the brown and the green cable.

Middle contact in lower position
In this position the resistance is tested, through which the current flows at maximum board voltage when the magnet develops so much force that it pulls the middle contact against the lower one.

Figure 9-38

A resistance of 8.4 ohms should now be measured between the brown and the green cable

If the measured resistance values correspond to the desired values, the voltage regulator is electrically okay. This does not necessarily mean that the voltage regulator will work properly.

Depending on the demand of the consumers and the engine speed the middle contact changes its position between the upper contact and lower contact in fractions of a second. The actual position of the middle contact depends on the interplay between the spring, which pulls the middle contact against the upper contact and the force of the magnet which pulls the middle contact against the lower contact. When the spring force is decreased by wear, the magnet pulls the middle contact too early down, that means, when the voltage of the vehicle electrical system is too low. This can be determined by measuring the voltage at the carbon brushes while the engine is running. It should be up to about 2000 rpm equal to the battery voltage and only then get lower, when the engine speed increases over 2000 rpm. The voltage measured at the battery should be about 14.5 to 15 volts at 2500 rpm.

However, you should only change the setting of the pretension of the spring, until you are sure that all other components work properly. Often the reason for a low charging voltage at the battery are oxidized contacts of plug connectors or in the ignition lock.

9.3.9 Testing the ignition lock

The entire current in the vehicle electrical system flows through the ignition switch. A malfunction of the ignition lock can therefore cause trouble everywhere in the electrical system. I describe the function test of the ignition lock as I used it in the simplified circuit diagram. The ignition lock has an inlet (the red cable) for the current from the battery. In the second key position, the current is passed to the brown cable (electrical system, switched plus). In the third key position the current is passed to the blue cable and the driving lights are switched on.

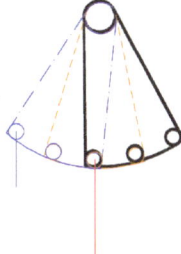

Figure 9-39 (circuit symbol)

The following defects may occur:

If the contacts in the ignition switch are oxidized, they represent a resistance for the current, which converts a part of the electric power of the generator into heat. This part of the electric power is thus no longer available for the ignition, the consumers and to charge battery.

Check the resistance with an ohm meter connected one after the other between the red and the brown and the blue cable in the corresponding key position. The measured resistance should be as small as possible, in no case more than 1 ohm.

A loose contact in the ignition lock can be responsible for misfiring, for which you may search in vain in the further ignition circuit. Such a loose contact may occur, for example, only at certain engine speeds. If you have to deal with unexplainable ignition problems, you should connect the orange cables of the ignition coils directly to batterie positive pole on and check whether the misfires still occur. In any case, the ignition coils must be disconnected immediately from the battery, as otherwise a current will flow constantly through one of the ignition coils, causing damage to the ignition coil and discharging the battery.

9.4 *The ignition circuit*

The design of the ignition circuit is quite simple, as shown in Figure 9-40: From the battery positive pole, a cable goes to the ignition switch and then to the two orange cables of the ignition coils. The brown cable of each ignition coil leads to the loose part of the respective contact breaker pair and to the capacitor. The fixed part of each pair of contacts is connected to the vehicle ground, just like the housing of the capacitor, thus closing the circuit. The third, central connection of the ignition coil is connected via the ignition cable to the spark plug, which is also connected to ground.

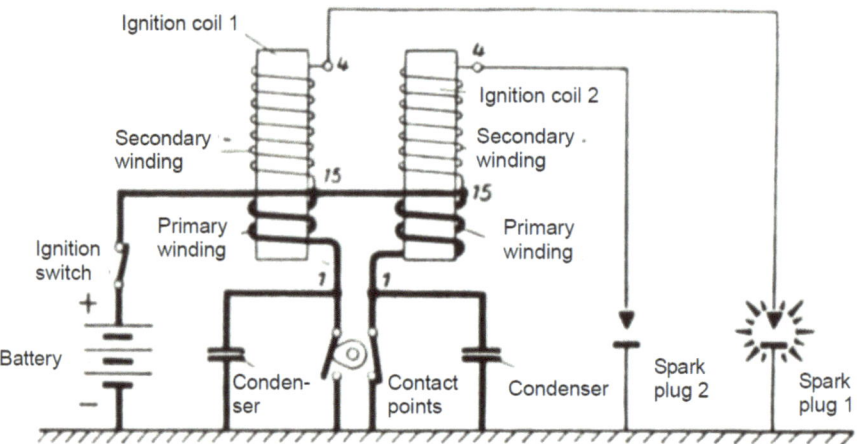

Figure 9-40

It would exceed by far the topic of this book to describe in detail all the components of the ignition circuit. On the following pages, however, I will try to describe the components as far that everyone should be able to install the ignition circuit, set the ignition and detect any faults.

A defective ignition coil can not be detected by means which are usually available. The same applies to the capacitor. Here, it only helps to exchange the corresponding component against another, hopefully functioning. If the problem is fixed then, this component was the reason. The "heart" of the ignition is the contact breaker base plate with the ignition contacts for the right and left cylinder, placed under the small chrome-plated cover in the cylinder head on the left side.

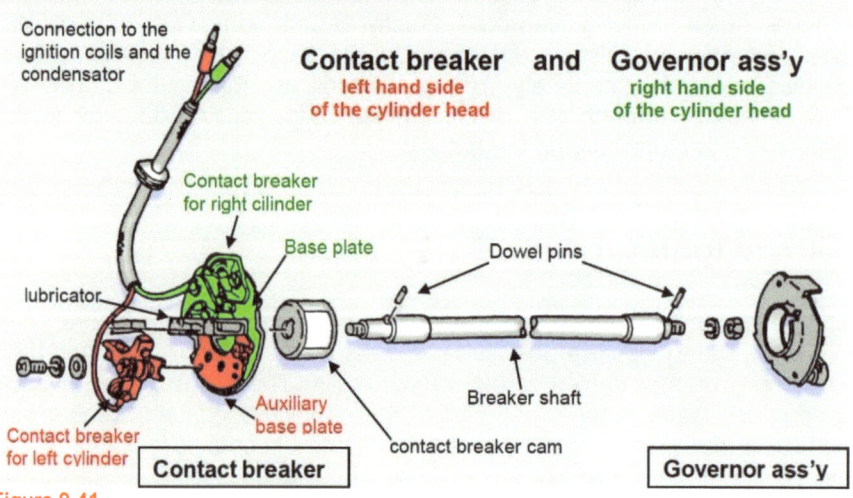

Figure 9-41

For the right and the left cylinder, I used marking colors: green (right) and red (left). The base plate (green) with the contact breaker pair for the right cylinder is bolted directly in the cylinder head, while the auxiliary base plate (red) with contact breaker pair for the left cylinder is bolted to the base plate. Therefore, the pair of contacts for the right cylinder must always be adjusted first. Both plates have arch shaped elongated holes, which enable the plates to be turned by a few degrees.

 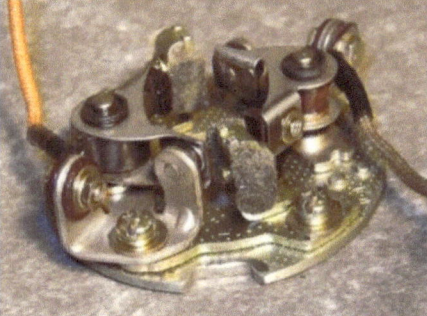

Figure 9-42

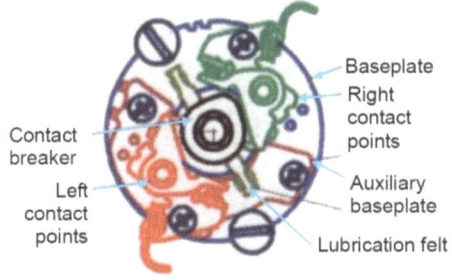 

Figure 9-43

How to adjust the ignition timing is explained later. The contact breaker pairs themselves are wear parts, which must be adjusted regularly and replaced after longer intervals. The red and green color markings in figure 9-43 have no meaning, both pairs are the same. However, the cables leading to the contact breaker pairs should also be marked in color, as shown in Fig. 9-42. Both pistons move up and down at the same time, one of which is in the compression stroke and the other in the exhaust stroke. If the connection cables from the contact breaker pairs to the ignition coils are mixed up, the ignition spark fires into the exhaust stroke. Of course the engine doesn't start and there are loud misfires.

For each contact breaker pair, there is one capacitor. The capacitors of the XS 650 are assembled together in one housing (figure 9-44). Here nothing can be mixed up - the connecting cables are connected to a brown cable of the ignition coils.

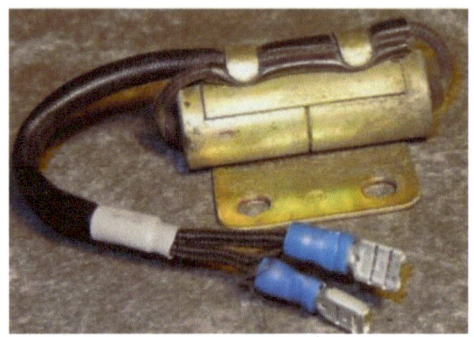

Figure 9-44

To generate a spark, a very high voltage of about 12 to 20,000 volts is required. Since a dc current present in the vehicle network can not be transformed, another effect is used. As already explained in chapter "9.2.1 The generator" a voltage is generated in an electrical conductor which is moved through a magnetic field. In case of the ignition, not the magnetic field or the electrical conductor is moved, but the voltage in the ignition coil changes.

Figure 9-45 shows the basic structure of an ignition coil.

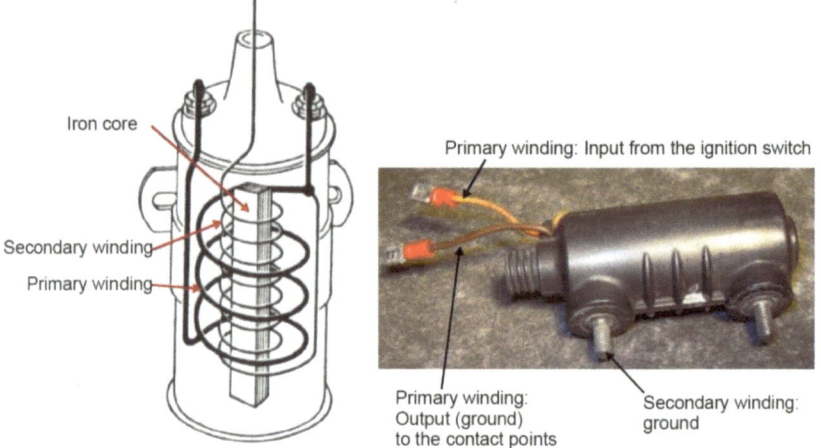

Figure 9-45

It consists of two coils around one iron core and it is basically designed like a transformer. If an ac voltage is applied to the coil with the few windings of thick wire (primary winding), a higher alternating voltage would emerge at the coil with many windings of thin wire (secondary coil). The ratio of the voltages depends on the ratio of the number of windings. In the case of a dc voltage, only a magnetic field is generated from the current-carrying coil with a few windings of thick wire, in which also the other coil is located. If the current in the first coil is interrupted, this magnetic field abruptly breaks down. This very fast change of the magnetic field results in a very high voltage in the secondary coil, which is discharged at the spark plug by an ignition spark. The current in the primary coil is interrupted by the contact breaker pairs, and the spark is triggered precisely at the moment when the contact breaker pair opens.The capacitor has the task of preventing sparking between the contact breaker pairs for the respective cylinders and storing energy. Burnt contact surfaces therefore indicate a defective capacitor.The contact breaker must open exactly at the moment when the piston is in the compression stroke just before the top dead center. When adjusting the ignition, the piston must be brought precisely into the position in which the spark is to be triggered. Therefore a marking is present on the rotor, which must be in line with a further marking on the crankcase cover.

Figure 9-46

In this position, the contact breaker pair of the right cylinder (colored green) should just begin to open. If this applies to the contact breaker pair of the left cylinder (red color), the crankshaft is rotated by 360 ° in the direction of rotation of the engine. Now you can start with the adjustment of the contact breaker pair for the right cylinder. How this is done is described step by step in the following chapter. Since the fuel / air mixture requires a certain time to burn completely, the ignition must be triggered earlier with increasing engine speed. Therefore, the interruption cam is turned against its direction of rotation by the governor (figure 9-47), so that the contact breakers open earlier. The connection between the governor and the contact breaker base plate can be seen in figure 9-41.

Figure 9-47

The governor cannot be adjusted, but it should be ensured that it is clean and smooth-running and that the weights return to their rest position when the engine is stopped. If this is not the case, the ignition is triggered too early and the kickstarter kicks back, what can lead to injuries. Due to wear of the mechanics, the ignition timing shifts to a later stage, which can not be corrected by the possibility of adjustment in the elongated holes of the base plates. The return springs of the weights of the governor can also get slack, which will cause, that the ignition is triggered earlier.

You can check this with a stroboscope gauge. I am assuming that the person who owns such a device is also familiar with the handling, and therefore I will not explain it in more detail.

At some point, both the base plates (worn out threads) and also the governor will be irreparably worn out and then you should think about an electronic ignition. With the electronic ignition, the mechanical governor is eliminated and maintenance-free sensors are installed instead of the contact breaker pairs. However, in the case of a possible defect, one can only exchange the control box or the sensors as a whole.

With the electronic ignition, there is the advantage that nothing has to be adjusted. So there is no irregular engine running and power losses due to non-synchronously set ignition times.

9.5 Adjusting the ignition timing

1. Set the contact distance to the right
Rotate breaker cam (at crankshaft stump) until maximum distance is reached (highest point of the cam). Loosen the fastening bolt and move the fixed part of the pair of contact breakers so that a distance of 0.3 to 0.4 mm is created between the contacts. Tighten the fastening bolt.

Figure 9-48

2. Set the contact distance to the left
Repeat procedure for left pair of contact breakers.

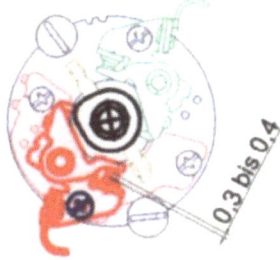

Figure 9-49

3. Set the ignition timing to the right
Turn the crankshaft until the ignition marking of the rotor is in line with the marking on the housing. Loosen the fastening bolts of the base plate. Turn the base plate until contact breaker pair for right cylinder is just opening (use a test lamp or continuity tester). In this position, tighten the fastening bolts.

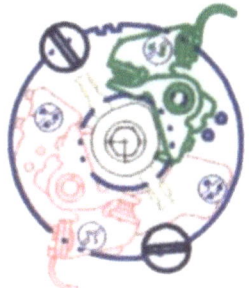

Figure 9-50

4. Set ignition timing to the left
Turn the crankshaft by almost 360° until the ignition marking of the rotor is again in line with the marking on the housing. Loosen the fastening bolts of the auxiliary base plate. Turn the auxiliary base plate until the contact breaker pair for left cylinder is just opening (use a test lamp or continuity tester). In this position, tighten the fastening bolts.

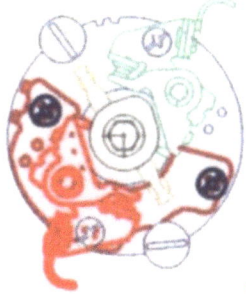

Figure 9-51

ABOUT THE AUTHOR

Hans J. Pahl is a German automotive engineer, who worked a couple of years as a design engineer with a medium sized company, which specialized in special purpose vehicles – mainly heavy duty trucks for oil fields and airfield fire engins. He worked 15 years in the R&D department and as a product developer with an international company in the automotive supply industry. Since 2002 he works as traffic accident investigator and as an expert for vehicle technology.

Hans J. Pahl has been riding motorcycles for 45 years now and since more than 20 years he owns a Yamaha XS 650 with a mileage of about 200.000 miles now.